CIGRE Green Books

T0172176

Technical Brochures

Series editor

CIGRE, International Council on Large Electric Systems, Paris, France

CIGRE presents their expertise in unique reference books on electrical power networks. These books are of a self-contained handbook character covering the entire knowledge of the subject within power engineering. The books are created by CIGRE experts within their study committees and are recognized by the engineering community as the top reference books in their fields.

More information about this series at http://www.springer.com/series/15774

Giorgio Diana
Editor

Modelling of Vibrations of Overhead Line Conductors

Assessment of the Technology

Editor
Giorgio Diana
Department of Mechanical Engineering
Politecnico di Milano
Milan
Italy

ISSN 2367-2625   ISSN 2367-2633   (electronic)
CIGRE Green Books
ISSN 2522-512X   ISSN 2522-5138   (electronic)
Technical Brochures
ISBN 978-3-030-10269-2   ISBN 978-3-319-72808-7   (eBook)
https://doi.org/10.1007/978-3-319-72808-7

Printed on acid-free paper

This Springer imprint is published by Springer Nature
The registered company is Springer International Publishing AG
The registered company address is: Gewerbestrasse 11, 6330 Cham, Switzerland

Foreword

Overhead Lines play an important role worldwide for the supply of electricity, and electricity will become even more important in the future.

Their conductors are responsible for the transport of energy and therefore need to be designed, constructed and mounted in a reliable, safe and long lasting manner. They have to fulfill their duties over decades with minimum efforts for maintenance.

Transmission lines are exposed to all kind of weather and environment and, as such, must be designed and protected adequately. Conductors and bundled conductors being probably the most flexible structures used on such large scale are particularly prone to vibrations. Consequently, it is extremely important to assess their severity of vibration to avoid conductor and accessories fatigue which may lead to failure. This is possible on laboratory spans, test lines or by measuring directly on existing lines. However, such tests and measurements take time and are expensive. Therefore, modelling becomes an extremely useful tool if it allows to predict aeolian vibration and subspan oscillation amplitudes at the design stage or to understand what is happening when a problem occurs on a line. It is also useful to predict the efficiency of damping systems and determine the optimal position of dampers, spacers and/or spacer dampers.

Conductor vibration amplitudes are modeled as the result of a balance between power imparted by the wind and power dissipated by conductor self-damping and damping devices. Those different aspects are covered in detail in this brochure. The work reported here aims at validating the accuracy of existing models to perform such calculations. It collects five papers already published in ELECTRA and in CIGRE Science & Engineering and makes them available in one document. The following topics are covered:

- Modelling of aeolian vibrations of single conductors;
- Modelling of aeolian vibrations of single conductors plus damper;
- Modelling of aeolian vibrations of single conductors strung at relatively high tensile load;

- Modelling of aeolian vibrations of bundled conductors;
- Modelling of subspan oscillations of bundled conductors.

We are convinced that this publications related to conductors is an important and valuable tool for designers of overhead lines and will also help for maintenance considerations and would like to express our sincere thanks to the editor and the authors for all their efforts.

Montreal, Canada   Pierre Van Dyke
Vienna, Austria   CIGRE TAG B2.06 Convenor
                                                                                   Herbert Lugschitz
                                                                       Chairman, CIGRE SC B2

Preface

Wind-induced vibrations of conductors must be controlled below critical levels to avoid fatigue damage and obtain reliable transmission lines.

Newly designed lines, lines that are being modified to carry higher current and/or voltage and lines that are being assessed for life extension, all require vibration control for safe levels.

Approaches available to guide this assessment process can be pragmatic through design rules based on the past experience. Also, conditions can be assessed through measurement on existing lines or test lines using special purpose measuring instruments.

However, such tests and measurements take time and are expensive. Therefore, modelling becomes an extremely useful tool which allows to predict Aeolian vibration and subspan oscillation amplitudes at the design stage or to understand what is happening when a problem occurs on a line. Numerical models are also useful to predict the efficiency of damping systems and determine the optimal position of dampers, spacers and/or spacer dampers, at the design stage.

The above is the spirit that has guided the work performed in about 20 years by different groups of experts and researchers inside Cigré Working Groups. This work resulted in five publications—the first in Electra, dating back to 1998, the last in the Cigre Science & Engineering Review (Vol. 2, June 2015), all of them dealing with the assessment of the technology available for wind induced conductors vibrations modelling.

They are now available all together in this book.

Many experts were involved in the studies and discussions: their names are reported at the beginning of each chapter of the book and I would like to thank all of them for the precious contribution given in sharing experience, contributing to discussion and writing of the different parts of the papers. A special and affectionate thought goes to my dear friend Chuck Rawlins, who is no longer with us but gave an outstanding contribution to all the work with its wide knowledge and experience.

Thanks are of course extended to the reviewers of the different chapters, for their diligence and useful suggestions.

I would also like to thank the different Convenors of the WG B2-11, under which the initial part of the work has been developed, for their suggestions and general guidance: Dave Havard, Konstantin Papailiou, Dave Hearnshaw.

Final thanks go to Pierre Van Dyke, the Convenor of the B2-AG 06 Group, which is the umbrella under which Working Groups dealing with the final part of the work have operated and Herbert Lugschitz, the present Chairman of Cigré Study Committee B2 (Overhead Lines).

Pierre also took active part in the work and in the preparation of the book.

Very final thanks to Konstantin Papailiou: dear tireless Costantino, a powerhouse of ideas, it is thanks to your efforts that this initiative could be finalized.

Milan, Italy   Prof. Giorgio Diana

Message from the Secretary General

Dear Readers,

Three years ago, I had the pleasure to comment on the launching of a new CIGRE publication collection, in an introductory message about the first CIGRE Green Book, the one on Overhead Lines.

The idea to evaluate the collective work of the Study Committees accumulated over more than twenty years, by putting together all the Technical Brochures of a given field, in a single book, was first proposed by Dr. Konstantin PAPAILIOU to the Technical Committee (now Council) in 2011.

One year later in 2015, the cooperation with SPRINGER allowed CIGRE to publish it again as a 'Major Reference Work' distributed through the vast network of this well-known international publisher.

Last year in 2016, the collection was enriched with a new category of Green Books, the CIGRE 'Compact Series', to satisfy the needs of the Study Committees when they want to publish shorter, concise volumes.

The first CIGRE Compact Book was prepared by Study Committee D2, under the title 'Utility communication networks and services'.

Since then the concept of the CIGRE Green Books series has continued to evolve, and today you have in your hands or on your screen, the first volume of the third subcategory of the series, the 'CIGRE Green Book Technical Brochures' (GBTB).

CIGRE has published more than 700 Technical Brochures since 1969, and it is interesting to note that in the first one, on Tele-protection, the first reference was a SPRINGER publication of 1963.

A CIGRE Technical Brochure is produced by a CIGRE Working Group, following specific Terms of Reference, is published by CIGRE Central Office and available from the CIGRE on-line library, e-cigre, one of the most comprehensive, accessible databases of relevant technical literature on power engineering.

Between 40 and 50 new Technical Brochures are published yearly, and these brochures are announced in Electra, CIGRE's bimonthly journal, and are available for downloading from e-cigre.

From now on, the Technical Council of CIGRE may decide to publish a Technical Brochure as a Green Book in addition to the traditional CIGRE Technical Brochure. The motivation of the Technical Council to make such a decision is in order to disseminate the related information beyond the CIGRE community, through the SPRINGER network.

As the other publications of the CIGRE Green Books series, the GBTB will be available from e-cigre in electronic format free of charge for CIGRE members.

CIGRE plans to co-publish new Green Books edited by the different Study Committees, and the series will grow progressively at a pace of about one or two volumes per year.

In a few final words I would like to thank and congratulate all the authors, contributors and reviewers of this specific publication on the 'Modelling of Conductor Vibrations'.

Paris, France   Philippe ADAM
                                                                      Secretary General

Executive Summary

Modelling of Aeolian Vibrations of Single Conductors

This chapter deals with an analytical approach which may be used to investigate alternatives in the design or redesign process of a line. In particular, this section describes the energy balance principle (EBP) which is used to estimate an upper bound to the expected vibratory motions, gives examples of measured wind and conductor self-damping data used, and some comparisons with available field measurements.

Important problems are tackled related to the reliability of available data, investigated through:

1. a critical analysis of the available data pertaining to wind power input;
2. an analysis of data on conductor self-damping;
3. comparison of analytical and experimental vibration measurements;
4. study of how uncertainties regarding wind power input and conductor self-damping are reflected in analytical predictions of vibration behaviour, obtained through the EBP (Energy Balance Principal).

These results offer guidance for future research to improve the reliability of such predictions.

The main conclusion is that through predictions of aeolian vibration level in operating lines, obtained using the EBP and the various available databases, it is possible to obtain a good reproduction of the frequency range and of the distribution of vibration amplitudes with frequency.

If the wind power functions and self-damping models employed in the study are indicative of the range of uncertainty in these parameters, then the range of uncertainty in EBP predictions of vibration amplitude can be about ± 50–60%, when steady-state conditions aren't reached.

Modelling of Aeolian Vibrations of Single Conductors Plus Damper

It is well known that if the conductor tension (or, more precisely, the ratio between tension and conductor unit weight H/w) exceeds certain limit values, aeolian vibrations may cause serious conductor and fitting damage. This limit H/w value is generally exceeded on transmission lines and then it is established practice to protect conductors with suitable dampers.

For new transmission line designs, it is important to know how much additional damping is needed to control aeolian vibration within safe levels. To this purpose, various researchers have developed computation methods, based on the EBP, to predict the aeolian vibration level of a conductor plus damper and then to allow for the selection of the suitable damping system: these methods use the damper dynamic characteristics as measured on a shaker.

This section evaluates the computation methods, through direct comparison among them and with results obtained on an experimental span, with the final aim of defining the uncertainty of the considered methodology.

The strains predicted by the different researchers exhibit considerable variability. Nevertheless, analytical methods based on the EBP and shaker-based technology can provide a useful tool for use in design of damping systems for the protection of single conductors against aeolian vibrations. It should be used with circumspection and be supplemented by references to field experience.

Modelling of Aeolian Vibrations of Single Conductors Strung at Relatively High Tensile Load

It has been shown that analytical methods based on the EBP and a shaker-based technology can provide a useful design tool for damping systems that protect a single conductor against aeolian vibration.

It is then important to evaluate the effectiveness of these methods for the design and/or verification of the damping system of long, single conductor spans strung at relatively high tensile load, such as crossings, which need more than one damper per span extremity to be effectively damped against aeolian vibration.

As in the first two sections, this one is based on an analysis of the available technology and on the results of two benchmarks: an analytical–analytical benchmark and an analytical–experimental one. The comparison between the analytical results produced by the different available models and the experimental one helped to understand the limitations and the usefulness of the approach.

The WG focused its attentions on long spans that are more critical than standard length spans discussed in the two previous sections. In fact, the longer the span length, the more unrealistic is a constant wind profile along the line, especially at the low wind speeds required to produce aeolian vibration. Moreover, an additional parameter which becomes significant for large sags is the tensile load variation along the span. It affects the conductor vibration wavelength.

In the work, it was demonstrated that, the use of the EBP approach, i.e. the assumption of a constant wind speed along the span, for long span applications should guarantee predicted vibration amplitudes higher than those that occur in reality, therefore producing conservative damping system designs.

However, future research work is needed:

- to improve the EBP technology, which generally produces a safe design of the damping system;
- to better understand the effect of turbulence and mean wind speed variation;
- to better simulate the mechanical system, in such a way to reproduce tensile load variations and multi-frequency excitation.

Modelling of Aeolian Vibrations of Bundled Conductors

This section covers the effectiveness of analytical methods based on the EBP and a shaker-based technology for the design and/or verification of the damping system of conductor bundle spans with respect to aeolian vibrations.

This study is based on an analysis of the available technology and on the results of benchmarks: an analytical–analytical benchmark and an analytical–experimental one are used for the evaluation relevant to aeolian vibrations.

The two benchmarks assessed within the Working Group allowed to understand the main differences among the models presently adopted in the field of conductor vibrations to understand and control aeolian vibration.

The analytical–analytical benchmark showed that the computed vibration amplitudes have a very similar trend even if some differences in the models are present.

Regarding the experimental-numerical benchmark, the numerical results generally exceed the experimental ones and then they are conservative, at least at low frequencies. However, it must be pointed out that, generally, when dealing with twin bundles, numerical results appear to be less conservative in respect to the experimental data.

The sensitivity analysis demonstrated that a non-negligible influence in the assessment of conductor behaviour, when dealing with aeolian vibrations, is given by the introduction of tension differentials and variable wind turbulence with wind speed.

Clearly, it is not straightforward knowing the real value to assign to the turbulence and to the tension differentials when the bundle behaviour for aeolian vibrations must be analysed.

Future work is needed to achieve better knowledge on this issue, considering the comparison between measurements on a real line and analytical results.

Modelling of Subspan Oscillations of Bundled Conductors

Subspan oscillations are a well-known phenomenon in high voltage and ultra high voltage overhead transmission lines (HV and UHV OHTL). It occurs on conductor bundles and it is due to the effect of the wake produced by the windward conductor on the leeward one. For this reason, the phenomenon is also classified as wake induced oscillations: this phenomenon is a flutter type instability due to the coupling of vertical and horizontal modes in a frequency range between 0.5 and 2 Hz. It may lead to conductor failure in the spacer clamp (see Fig. 1) or suspension clamps or to spacer damper articulation failure.

Recently, problems associated with this phenomenon have become more recurrent, attracting the attention of transmission line operators, hence WG B2.46 decided to evaluate the developments in this field.

**Severely damaged conductor under
spacer clamp due to subspan oscillation.**

Fig. 1 Damaged conductor after subspan oscillation

In the past, several simulation models have been developed. In these models, the motion of the leeward cylinder is studied along two orthogonal directions, the windward cylinder being still. Then, the linearized quasi-steady theory (in the following QST) is employed and the drag and lift coefficients on the leeward cylinder are deduced from static measurements in wind tunnel, as a function of the relative position of the leeward cylinder with respect to the windward one (Fig. 2).

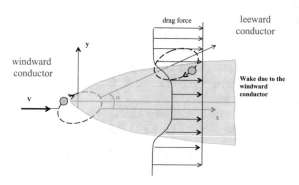

Fig. 2 Schematisation of subspan mechanism modelling

Such models are linear and clearly simplify the structural behaviour of the bundle subconductors, taking it back to a two degrees of freedom system in which the leeward conductor is the only one moving.

Nowadays, the finite element model (in the following FEM) analysis allows for the reproduction of the bundle dynamics and for the application of the aerodynamic forces to each subconductor using the QST with a nonlinear approach.

However, FEM analyses in the time domain are not always a practical tool for subspan oscillation simulation also because of the computation time required to obtain results.

All the models developed up to now are based on the QST: according to this, the field of forces in the wake of the windward conductors is accounted for using static aerodynamic coefficients measured in the wind tunnel and the effect of relative motion between subconductors corresponds to a relative velocity with respect to the approaching flow.

Another important issue faced in the work is the Reynolds number (Re) effect on the phenomenon.

In fact, for stranded conductors, i.e. rough cylinders, with the typical values of conductor diameter and wind speed involved by subspan oscillations, Re may be close to the critical zone: hence the Re number could significantly affect the phenomenon, due to the non-negligible variations of the drag coefficient with Re itself.

The study presents several approaches to the evaluation of the subspan phenomenon, ranging from approaches based on the EBP to approaches relying on FEM modelling.

In the work, a benchmark within the different type of models available for subspan oscillation studies is carried out comparing numerical results with measurements on the IREQ Varennes test line equipped with a quad bundle of ACSR Bersimis conductors and spacer dampers.

The obtained results allow to state that:

- the QST seems able to well reproduce the aerodynamic forces produced during subspan oscillations;
- the Reynolds number affects in a large amount the energy introduced by the wind;
- the numerical model based on EBP approach and on sophisticated wind tunnel tests to identify the aerodynamic parameters seems to be a useful tool for analysing the subspan oscillations phenomenon.

This document is highlighting what can be expected from numerical models regarding conductor vibrations.

- assessment of the aeolian vibration condition of particular lines, with conductors whose mechanical properties are poorly defined, or with special terrain conditions, may require field measurements;
- analytical methods based on the EBP and shaker-based technology can provide a useful tool to design damping systems for the protection of single conductors against aeolian vibrations;
- future research work is needed to improve the EBP technology, which generally produces a safe design of the damping system, in order to provide reliable results on long spans.

This document is reporting on the state of the art regarding aeolian vibrations and subspan oscillations modelling. Of course, this field of expertise is not static and research, numerical as well as experimental, is still going on in order to improve our knowledge of the phenomenon and translate it into improved numerical models.

Contents

About the Editor

Giorgio Diana received his Degree in Mechanical Engineering from Politecnico di Milano in 1961. Since 1971 he is a Full Professor of Applied Mechanics and in December 2010, he has been appointed as Professor Emeritus of Politecnico di Milano. His actual main research fields are: bridges dynamics, railway vehicle—infrastructure dynamic interaction and aero-elasticity problems.

He is presently chairing various scientific committees such as the Steering Committee of the Joint Research Centre (JRC) on Transportation, the Italcertifer Consortium Certification Committee, the Working Group on Super-long Span Bridge Aerodynamics of IABSE (International Association of Bridge Engineers) and the Working Group B2-58 (Vibration Modelling of High-Temperature Low Sag conductors—Self-damping characterization) of CIGRE (Conseil International des Grands Réseaux Electriques).

The results of the research and activities are reported in about 300 papers published mainly in international peer-reviewed journals as well as in keynote lectures held at several international conferences.

As a result of his achievements, Prof. Diana received in 2009 the Robert H. Scanlan Medal from the American Society of Civil Engineers (ASCE), in 2012 the CIGRE Technical Committee Award, in 2014 the IAWE Senior Award with Davenport medal and, also in 2014, the European Railway Award from the Community of European Railway and Infrastructure Companies (CER) and from the European Rail Industry Association (UNIFE).

List of Symbols and Abbreviations

EBP	Energy Balance Principle
H/w	ratio between cable tensile load and cable unit weight
HV	High Voltage
UHV	Ultra High Voltage
OHTL	Over Head Transmission Line
ACSR	Aluminium Conductor Steel Reinforced
AACSR	Aluminium Alloy Conductor Steel Reinforced
ACAR	Aluminium Conductor Alloy Reinforced
QST	Quasi Steady Theory
FEM	Finite Element Model
$\mathrm{Re} = \rho VD/\nu$	Reynolds' number
ρ	fluid density
ν	fluid viscosity
D	cable/conductor diameter
V and U	fluid/wind velocity
y_0/D	non-dimensional antinode vibration amplitude
y_0, u	antinode vibration amplitude (0-peak value)
P_w	wind power imparted to a unit length of conductor
f	vibration frequency
$\mathrm{fnc}(y_0/D)$	reduced power function
$\mathrm{St} = f_\mathrm{St}D/V$	Strouhal number
f_St	Strouhal frequency
$\mathrm{Sc} = \delta m/\rho D^2$	Scruton number
δ	logarithmic decrement
m	cable/conductor mass per unit length
PT	Power Method
ISWR	Inverse Standing Wave Ratio
P/L and W/L	Power dissipated per unit length of conductor
E	Energy dissipated per unit length of conductor
k, DC	factor of proportionality in the P/L and E empirical laws

l, m, n	amplitude, frequency and tension exponent in the P/L empirical law
T	conductor tensile load
UTS	Ultimate Tensile Strength
$I_t = V_{RMS}/V_{MEAN}$	Index of turbulence
RMS	Root Mean Square value
EJ_{max}	Conductor flexural stiffness computed with no interstrand slipping
EJ_{min}	Conductor flexural stiffness computed with total interstrand slipping
err_a, err_b	error factors
P_d/P_{max}	damping efficiency
$f\, y_{max}$	vibration intensity (product between vibration frequency and maximum antinode amplitude of vibration)
AA	Analytical-Analytical
AE	Analytical-Experimental
$V_r = V/(f\ I)$	Reduced Velocity
I	cylinders/conductors separation
I/D	ratio between bundle separation and conductor diameter
C_D	Drag coefficient
α	bundle rotation angle with respect to the wind
x/D	non-dimensional oscillation amplitude
$q_o(t)$, $q_v(t)$	modal coordinates (horizontal and vertical modes)
S_{am}, S_{im}	maximum amplitude of the horizontal (q_o) and vertical mode (q_v)
ω	circular frequency of the horizontal mode
2D	Two-Dimensional
3D	Three-Dimensional

Chapter 1
Modelling of Aeolian Vibrations of Single Conductors

G. Diana, F. Tavano, L. Cloutier, R. Claren, M. Ervik, P. Hagedorn, C. Hardy, G. Kern, H-J. Krispin, L. Möcks, C. B. Rawlins, P. W. Dulhunty, A. Manenti, M. Tunstall, J. M. Asselin, W. Bückner, D. G. Havard and D. Hearnshaw

1.1 Introduction

Reliable transmission line design requires that vibration of the conductors due to wind is controlled below critical levels to avoid fatigue damage. Newly designed lines, lines that are being modified to carry higher current voltage and lines that are being assessed for life extension, all require vibration control for safe levels. Approaches available to guide this assessment process can be pragmatic, through design rules based on the past experience. Also, conditions can be assessed through measurement on existing lines, using special purpose measuring instruments. This paper deals with an analytical approach which may be used to investigate alternatives in the design or redesign process. In particular, this paper describes the energy balance principle which is used to estimate an upper bound to the expected vibratory motions, gives examples of measured wind and conductor data used, and some comparisons with available field measurements.

As a first step in defining the dynamic behaviour of single conductors subjected to aeolian vibration, studies have been made of undamped single conductors exposed to low turbulence winds. This is the simplest case and existing wind tunnel data corresponds more closely to low turbulence winds. Such winds are encountered in very flat terrain, at the river and straits crossings, and during periods of thermal inversion over many other types of terrain.

Analysis of aeolian vibration behaviour of overhead conductors generally makes use of the so-called energy balance principle (EBP) (CIGRE SC22 WG01 1989). The EBP works in the frequency domain: in its simplest form, one mode of vibration at a time is considered and steady-state solutions computed correspond to

G. Diana (✉)
Department of Mechanical Engineering, Politecnico di Milano, Milan, Italy
e-mail: giorgio.diana@polimi.it

© Springer International Publishing AG 2018
G. Diana (ed.), *Modelling of Vibrations of Overhead Line Conductors*,
CIGRE Green Books, https://doi.org/10.1007/978-3-319-72808-7_1

the maximum vibration amplitude which could be excited on that conductor at that frequency. Transient effects like those due to wind turbulence cannot be taken into account. Some advanced analyzes take into account the statistical distribution of the wind on the conductor, either in the time (Diana et al. 1993) or the frequency domain (Noiseux et al. 1988; Rawlins 1983a). However, the straightforward EBP is considered acceptable for use for engineering applications and a number of computer programmes have been written around it, most of which can be implemented on a personal computer.

The reliability of results of these analytical computations is no better than the background data used in them, particularly, data on the power supplied by wind during aeolian vibration and data on self-damping in stranded conductors. This document concerns the reliability of those results.

To make this assessment, the study was done in four parts, as follows:

1. 'Laminar Wind Power Input' presents a critical analysis of the available data pertaining to wind power input.
2. 'Conductor Self-damping' presents the corresponding analysis of data on conductor self-damping.
3. 'Comparison between Analytical and Experimental Results', several of the available programmes based on the EBP are used to compare analytical with experimental vibration measurements.
4. 'Influence of Uncertainty about Self-Damping and Power from Wind on the Analytical Prediction of Conductor Response' reports results of a study of how uncertainties regarding wind power input and conductor self-damping are reflected in analytical predictions of vibration behaviour. These results offer guidance for future research to improve the reliability of such predictions.

1.2 Laminar Wind Power Input

Vortex shedding excitation is a very complex aeroelastic phenomenon. Much research has been carried out on the subject, not only in the air but also in water and other fluids.

If attention is focused on the energy introduced by a laminar wind to a vibrating conductor model, it is possible to define the maximum energy introduced by the wind to a conductor as a function of the nondimensional amplitude of vibration y_0/D (y_0 is the antinode amplitude of vibration and D is the conductor diameter).

The several empirical functions available have been obtained mainly by means of wind tunnel tests (rigid rods, or flexible models vibrated as sine waves).

Besides wind tunnel tests, there are also theoretical studies: for example those by T.O. Slethey.

Most of the empirical functions derived by wind tunnel tests can be brought into the form:

$$P_w = f^3 D^4 \text{fnc}(y_0/D), \tag{1.1}$$

where:

- P_w is the wind power imparted to a unit length of conductor
- f is the vibration frequency, which is related to wind velocity through the Strouhal formula
- fnc is the reduced power function.

Figure 1.1 illustrates some of these fnc(y_0/D) functions; the dispersion among the results collected by the various researchers may be due to several causes. These may be summarized as follows:

- Three-dimensional effects
  To avoid or, at least, minimize them, the incident flow should be two-dimensional. End discs should be provided at the model extremities or, better, only forces acting on a portion of the cylinder should be measured, the energy input being computed from them.
- Wind tunnel turbulence
- Cylinder roughness
- Reynolds' effect.

Generally, while it is possible to take into account the Strouhal parameter (St = $f_{St}D/V$, where f_{St} is the Strouhal frequency, D is the cylinder diameter and V is the wind velocity), the same is not true of Reynolds' number, which is defined as Re = $\rho VD/v$ (where ρ and v are the fluid density and viscosity). Reynolds number represents the ratio between inertia and viscous damping forces. Tests made with different Reynolds numbers give different results. In particular, as Reynolds number

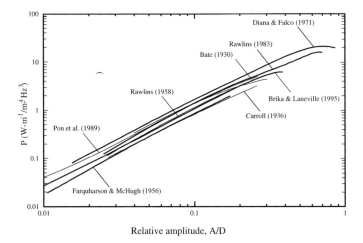

Fig. 1.1 A comparative study of the power imparted by wind to a circular cylinder vibrating in a sinusoidal shape versus nondimensional amplitude (Brika and Laneville 1995)

increases, the power input becomes smaller, and the phenomenon may be completely changed when critical or so-called transition Reynolds' numbers are approached.

All these effects result in a reduction of wind power imparted to the conductor.

Another cause of the dispersion evident in Fig. 1.1 may be the fact that some researchers use rigid rods to represent the conductor, while others use flexible models.

The differences in terms of maximum nondimensional amplitude y_0/D reached by the various researchers in their wind tunnel tests are due to different Scruton numbers used. Scruton number represents the ratio between damping and aerodynamic forces and is defined as $Sc = \delta m/\rho D^2$, where δ is the logarithmic decrement and m is the conductor mass per unit length.

It can be observed that Diana and Falco are the researchers who made wind tunnel tests with the lowest Scruton number, reaching the maximum vibration amplitudes ($y_0/D = 0.8$). Transmission line conductors can be characterized by very low Scruton numbers and vibration amplitudes can reach one diameter.

However, it seems unlikely that all of the dispersion shown in Fig. 1.1 can be explained on the basis of the above-mentioned effects.

In order to evaluate which of the different wind power input curves better fit reality, comparisons between experimental and theoretical results of aeolian vibration level can be performed. Theoretical results should be obtained by EBP, once the conductor damping behaviour has been assessed by means of laboratory tests. Two examples are given in the section, 'Comparison between experimental and theoretical results'.

1.3 Conductor Self-damping

Conductor self-damping (Hardy and Leblond 1993) designates a physical characteristic of the conductor which defines its capacity to dissipate energy internally while vibrating. For conventional stranded conductors, energy dissipation is thought to be due mainly to structural causes, i.e. to reciprocating frictional micro-slip within the multitude of tiny contact patches between overlapping individual wires, as the conductor flexes with the vibration wave shape.

This characteristic is important as it governs otherwise-undamped conductor response to vortex-induced excitation (aeolian vibrations) over much of the frequency range of interest. It thus determines the range of frequencies where vibration dampers may be needed. Methods for measuring conductor flexural self-damping have been specified in an IEEE Standard which came into force in 1978: Standard 563-1978 'IEEE Guide on Conductor Self-Damping Measurements' (see also CIGRE SC22 1979).

A collection of such measured self-damping of several conductors and OPGWs, originating from five different laboratories, is provided in Appendix A (This appendix is available on request at the Central Office).

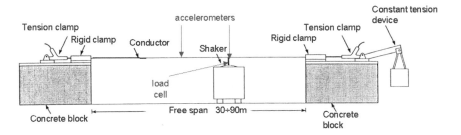

Fig. 1.2 Basic test span layout

Conductor self-damping is generally measured in a laboratory test span as sketched in Fig. 1.2.

The test span comprises two massive blocks, 30–90 m apart, onto which the conductor to be characterized, strung to the required tension, is rigidly held. The conductor is then brought successively into a sequence of resonance frequencies at a controlled antinode amplitude by means of an electromagnetic shaker.

Two methods have been used by most investigators in conductor vibration research.

One method, known as the power method (PT) consists in measuring the total power transmitted to the conductor at the connection point with the shaker. This power comprises the power dissipated by the conductor both in-span and at span ends through structural damping and aerodynamic damping, and the power dissipated in the terminations, if significant. Whenever the stiffness of the terminations is high enough, end-span damping is mostly due to structural damping within the conductor in a short segment near termination clamps where it is forced to bend severely. Span-end effects can be evaluated, as suggested by the IEEE Std, through comparison of measurements obtained with different test span lengths and/or they can be minimized by the use of pivoted clamps.

The second method, known as the inverse standing-wave ratio method (ISWR), is based on the measurement of the power flow through the conductor. It thus leads directly to the measurement of in-span damping, within free vibration loops, and, possibly, to end-span damping as a separate measurement.

Data measured in the laboratory span are generally expressed empirically through a power law:

$$\frac{P}{L} = k \frac{y_0^l f^m}{T^n},\qquad(1.2)$$

in which P/L designates power per unit length dissipated by the conductor; k is a factor of proportionality; y_0 is antinode displacement amplitude of vibration; f is the frequency of vibration and T is conductor tension. l, m and n are the amplitude exponent, frequency exponent and tension exponent, respectively.

Using the above empirical rule, self-damping determined in short laboratory spans can then be extrapolated to actual, much longer spans.

Table 1.1 summarizes the exponents reported by a number of investigators (CIGRE SC22 WG01 1989; Tompkins et al. 1956; Claren and Diana 1969; Seppä 1971; Rawlins 1983a, b, c; Noiseux 1992; Tavano 1991; Möcks and Schmidt 1989; Kraus and Hagedorn 1991; Cigada and Manenti 1996; Seveso 1996) together with the method of measurement used, the test span length and span-end conditions as well as number of conductors (cables) and tensions tested.

It will be noticed that the PT method for conductor self-damping measurements on laboratory test spans with rigidly fixed extremities produces empirical rules with an amplitude exponent close to 2.0 and a frequency exponent close to 4.0, in comparison to about 2.4–2.5 and 5.5, respectively, for the ISWR method and PT method with pivoted extremities.

Such differences in the above exponent values, together with those in the k factor of proportionality, may lead to huge differences in the predicted self-damping values.

This is shown typically in Figs. 1.3 and 1.4, where expected power/unit length dissipated by an ACSR Drake conductor strung at 30% UTS is plotted as a function of frequency for two different antinode velocities of vibration, 0.25 and 0.5 m/s, respectively, according to the rule proposed by three different investigators: exponents of rule (Claren and Diana 1969) and of rules (Seppä 1971; Noiseux 1992) represent extreme values of the range of variability above defined.

At the low frequencies, predicted self-damping according to Claren and Diana (1969) may exceed self-damping derived from Seppä (1971) and Noiseux (1992) by one order of magnitude. The disparity, however, attenuates with an increase in frequency. Obviously, curves derived from Seppä (1971) and Noiseux (1992) agree fairly well in the whole frequency range of interest.

In order to further explain disparities in self-damping prediction, a simulation was carried out on the basis of an ACSR Bersfort conductor strung at 30% UTS on 31.5 and 63 m long test spans, respectively (both with rigid extremities), by means of Noiseux' similarity laws related either to in-span (free-field) damping and end-span (near-field) damping. This discussion is based on a previous SC22 WG 11 report (Hardy and Leblond 1993). Results are shown in Figs. 1.5 and 1.6 which apply to tests at constant antinode velocities equal to 0.25 and 0.50 m/s, respectively. Five curves are plotted in each figure: the lower one corresponds to the in-span structural damping (i.e. total in-span damping free of aerodynamic damping and end effects) per unit length of the conductor. In-span structural damping does not depend upon span length. The second curve applies to total structural damping per unit length averaged over the span (in-span damping plus end-span damping) for the 63 m long span; the middle curve is the same, but with aerodynamic damping included, that is, the total power supplied by the shaker. The penultimate curve corresponds to total structural damping per unit length for the 31.5 m long span and, finally, the upper curve, to total damping, including aerodynamic, for the latter span.

Table 1.1 Comparison of conductor self-damping rules

Investigators	l	m	n	Method	End cond.	Span length (m)	No. of conductors $\times$ tensions
Tompkins et al. (1956)	2.45	5.5	1.9[a]	ISWR	NA	36	1×2
Claren and Diana (1969)	2.0	4.0	2.5; 3.0; 1.5	PT	MB	46	3×3
Seppä (1971) (Noiseux 1992)	2.5	5.75	2.8	ISWR	NA	36	1×8
Rawlins (1983b)	2.2	5.4		ISWR	NA	36	1×1
Lab. A (CIGRE SC22 WG01 1989)	2.0	4.0		PT	MB	46	1×1
Lab. B (CIGRE SC22 WG01 1989)	2.2	5.2		PT	PE	30	1×1
Lab. C (CIGRE SC22 WG01 1989)	2.44	5.5		ISWR	NA	36	1×1
Kraus and Hagedorn (1991)	2.47	5.36	2.8	PT	PE	30	$1 \times ?$
Noiseux (1992)[b]	2.44	5.63	2.76	ISWR	NA	63	7×4
Tavano (1991)	1.9–2.3	3.6–4.2		PT	MB	92	4×1
Mocks and Schmidt (1989)	2.45	5.38	2.4	PT	PE	30	16×3
Mech. Lab. Politecnico di Milano (Cigada and Manenti 1996; Seveso 1996; Diana et al. 2000)	2.43	5.5	2.0	ISWR	PE	46	4×2

ISWR: Inverse Standing Wave Method
PT: Power Method
NA: Not applicable
MB: Massive block
PE: Pivoted extremity
[a]Extrapolated
[b]Data corrected for aerodynamic damping

It may be appreciated that span end effects, in particular, are very significant. Obviously, the smaller the span length, the higher the contribution of the conductor ends to total damping. At the lower frequencies, damping is dominated by end effects and, to a lesser degree, by aerodynamic damping. The difference between total damping and in-span structural damping can easily reach one order of magnitude on short spans.

Now, the above power law may be written:

$$\frac{P}{L} = k_1 (y_0 f) f^{(m-1)}, \tag{1.3}$$

in which k_1 is another factor of proportionality. Hence, frequency dependence while testing at constant antinode velocity, $y_0 f = \text{constant}$, is

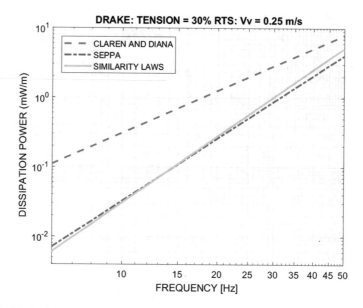

Fig. 1.3 Dissipation power per unit length versus frequency for ACSR Drake (antinode velocity 0.25 m/s)

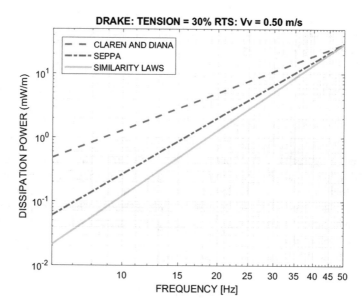

Fig. 1.4 Dissipation power per unit length versus frequency for ACSR Drake (antinode velocity 0.5 m/s)

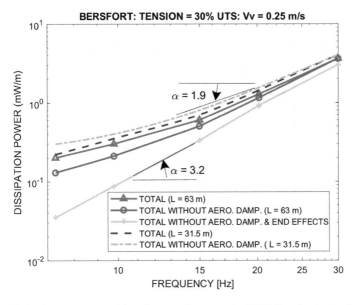

Fig. 1.5 Dissipation power per unit length versus frequency for ACSR Bersfort (antinode velocity 0.25 m/s)

$$\frac{P}{L} = k_2 f^{(m-1)} \tag{1.4}$$

The compound exponent $(m - 1)$ corresponds to the slope of the curves in Figs. 1.5 and 1.6. While curves related to total damping are not perfectly straight lines according to this simulation, the discrepancy is slight and it may be seen that average slopes are not far from 2.0 which, roughly, is the value derived from PT user's results (fixed extremities).

It thus appears that the major disparities amongst conductor self-damping values reported by different laboratories are mainly related to end effects. Therefore, the use of the PT for conductor self-damping measurement on laboratory test spans with rigidly fixed extremities is of questionable accuracy; use of pivoted extremities is suggested whenever this method is used.

1.4 Comparison Between Experimental and Theoretical Results

Two sets of experimental results were considered and, for each of them, various TF1 Members evaluated theoretically the aeolian vibration level by using their own data on wind power input and conductor self-damping.

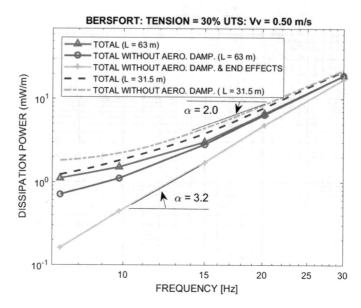

Fig. 1.6 Dissipation power per unit length versus frequency for ACSR Bersfort (antinode velocity 0.5 m/s)

The comparisons were made for data relevant to the antinode amplitude of vibration of the conductor.

As observed, computations and measurements refer to two different test sites whose spans were equipped with the conductors whose main characteristics are summarized in Table 1.2.

Figures 1.7 (Hardy and Van Dyke 1995) and 1.8 (Tavano 1991) give the measured turbulence level as a function of the average wind speed for the two test sites. Turbulence intensity is defined as the ratio between the standard deviation and the mean value of the wind speed $V(t)$. It is to be noted that this type of turbulence distribution is confirmed also by measurements performed by Kraus and Hagedorn (1991) as shown in Fig. 1.9.

When measured, level of turbulence is low, laminar (no turbulence) wind power input can be used for computations. In fact, as shown in Diana et al. (1993) using a more sophisticated approach based on the f.e.m conductor schematization and vortex shedding effects reproduced by means of equivalent nonlinear oscillators, the wind power input with low turbulence level is practically equivalent to that relevant to laminar wind condition.

The theoretical results are shown in Figs. 1.10 and 1.11, together with the experimental ones.

In both figures, four theoretical curves are reported: curve (1) refers to Diana and Falco (1971) no turbulence wind power input and to empirical conductor damping rule reported by Claren and Diana (1969); curve (2) refers to Diana and Falco (1971) no turbulence wind power input and to the semi-empirical conductor

Table 1.2 Conductors

Conductor type	ACAR 1300 (Diana et al. 1993)	ACSR
Material	ACAR	ACSR
Composition	19 Al/18 Alloy	54 Al/19 St
Diameter (mm)	33.33	31.5
Mass (kg/m)	1.816	1.92
Ultimate Tension (daN)	14,550	17,161
Every day Tension (daN)	2910	3450
Span length (m)	457	453
Duration of field tests	8 weeks	6 weeks

damping rule reported by Noiseux (1992); curves (3) and (4) are computed according to hypotheses reported by Rawlins (1992) and Hardy (1991), respectively.

In Fig. 1.11, there are two curves (4) according, respectively, to 10 and 20% turbulence levels (Noiseux et al. 1988).

As shown in the figures, the agreement between the experimental and theoretical results is very good in the case of the ACAR 1300 conductor. In contrast, some discrepancies appear for the ACSR conductor, probably due to some lack of experimental data, which were recorded only for 6 weeks.

As Fig. 1.10 shows, curves (2) and (4), which use the same empirical conductor self-damping rule, are quite similar; this is probably due to the fact that wind power

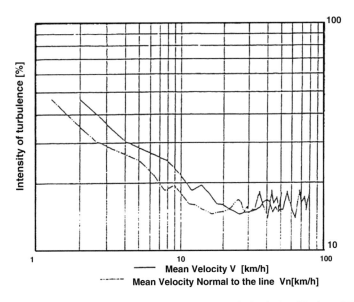

Fig. 1.7 First test site; intensity of turbulence versus mean wind velocity (Hardy and Van Dyke 1995)

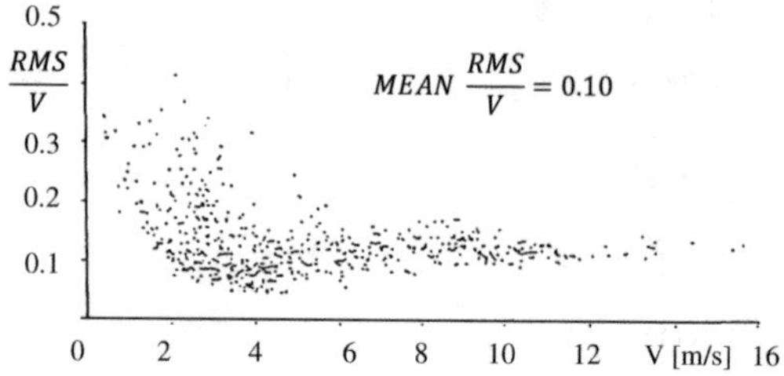

Fig. 1.8 Second test site; intensity of turbulence versus mean wind velocity (Tavano 1991)

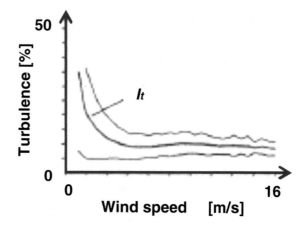

Fig. 1.9 Kraus and Hagedorn (1991) intensity of turbulence versus mean wind velocity

Fig. 1.10 ACAR 1300 Conductor max. value of antinode amplitude (m) versus frequency (Hz)

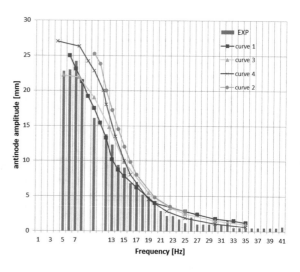

input functions are essentially the same. Considering now curves (4) and (1), it seems that in this case the two different assumptions of conductor self-damping, even if they represent extreme values in the range of variability of self-damping empirical rules (Table 1.1), do not compromise the experimental–theoretical

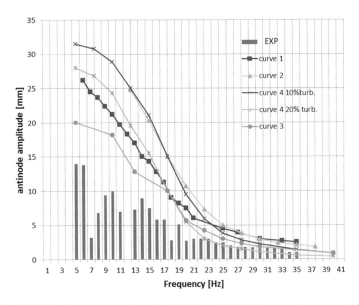

Fig. 1.11 ACSR Conductor max. value of antinode amplitude (m) versus frequency (Hz)

agreement. This fact can be explained by observing that the low-frequency maximum amplitude is not sensitive to the choice of the self-damping model because self-damping is small enough to be overshadowed by the variation of wind power with an amplitude at these frequencies.

The above illustrated predictions of aeolian vibration level in operating lines, obtained using the EBP and variously available databases, show that it is possible to obtain a good reproduction of the frequency range and of the distribution of vibration amplitudes with frequency.

The predicted amplitude level in some cases is quite close to the measured (Fig. 1.10), but in some other cases is quite different (Fig. 1.11). As already mentioned, discrepancies could be due also to some lack of experimental data and, even if in a smaller amount, to some imprecision in the data used for the EBP computations.

1.5 Influence of Uncertainty About Self-damping and Power from Wind on the Analytical Prediction of Conductor Response

The purpose of this section is to demonstrate the effect of uncertainty concerning power supplied by the wind and conductor self-damping upon the accuracy of predictions of aeolian vibration amplitudes, based on the EBP.

The range of uncertainty for wind power input can be observed in Fig. 1.1, which shows power curves from wind tunnel tests from various sources. The lowest curve is that of Carroll (1956), and the highest is that of Diana and Falco (1971). These two curves are shown in Fig. 1.12, together with a curve placed midway between them. A polynomial expression which fits this middle curve is $W[Wm^{-1}/(m^4Hz^3)] = -99.73\ (u/D)^3 + 101.62\ (u/D)^2 + 0.1627\ (u/D) + 0.2256$ (W is power imparted by the wind divided by span length, fourth power of conductor diameter and third power of frequency). Note that all of the recent curves in Fig. 1.1 fall above the middle curve of Fig. 1.12. Thus, the range of present uncertainty is taken to lie between the middle and upper curves of that figure.

In the case of conductor self-damping, the range of uncertainty is rather wide. Uncertainty comes, as already observed, from the measurement method and the conductor preconditioning but also from manufacturing procedure and control (Palazuelos et al.). In particular, it has been observed that there is a significant difference between results of measurements that were made using the PT on test spans with rigidly fixed extremities on the one side, and those made using the SWR or the PT with pivoted extremities on the other. The measurements of Claren and Diana (1969) are considered representative of the first group, and we take those of Noiseux (1992) to be representative of the second group. It has been seen that these two formulations differ significantly, especially at the low frequencies. Moreover, the Claren and Diana (1969) exponents represent the greatest departures from values based on the measurements of other investigators (see Table 1.1).

We note that recent measurements at the Mechanical Laboratory of Politecnico di Milano using both the PT and ISWR methods (Cigada and Manenti 1996; Seveso 1996), and taking effects of end-point damping into account, led to exponents (see Table 1.1) similar to those of Noiseux, although some differences remain. The difference between these two models was taken as a measure of the uncertainty associated with conductor self-damping.

The sensitivity of predicted amplitude to uncertainties about wind power and self-damping were assessed by calculating predicted free-loop amplitudes for three representative conductors, each at two tension levels, corresponding to values of the parameter H/w equal to 1557 and 2002 m, respectively (H is conductor tension and w is conductor weight per unit length) The calculations were made based on various combinations of the high and the middle wind power functions of Fig. 1.12, with the self-damping models of Noiseux (1992) and of Politecnico di Milano (Cigada and Manenti 1996; Seveso 1996; Diana et al. 2000).

Results of the calculations are presented as predicted free-loop amplitude versus frequency, as the vibration strain calculated from that amplitude and as the per-centage difference between predicted amplitude and the maximum of the several predicted amplitudes. They are reported in Figs. 1.13, 1.14, 1.15 and 1.16. The results point to several observations.

- In all cases, the amplitudes indicate vibration strain levels that may be con-sidered large enough to cause eventual fatigue.

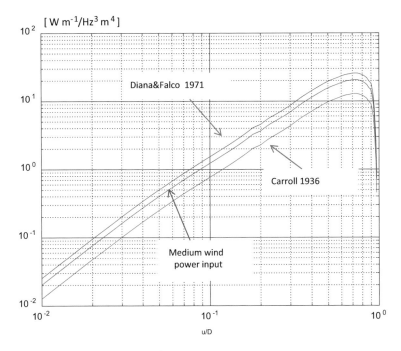

Fig. 1.12 Power imparted by the wind

- Maximum free-loop amplitude is around one conductor diameter for all combinations of assumed power input function and self-damping model and it occurs at low vibration frequencies. This low-frequency maximum amplitude is not sensitive to the choice of the self-damping model because self-damping is small enough to be over-shadowed by the variation of wind power with an amplitude at these frequencies. This amplitude is also not sensitive to the choice of wind power function, because both functions assumed in the study cross zero power at the same amplitude; about one diameter.
- If the upper limit of the frequency range of interest for aeolian vibration is taken as the frequency where calculated vibration strain falls below ±30 microstrain, it appears that it is the same for both self-damping models.
- For the lower H/w tension values, predicted vibration amplitude is more sensitive to the choice of wind power function (maximum variation between functions around 55%) than to the choice of self-damping model (maximum variation between models of around 25%). At the higher H/w level, the maximum variations were about the same, around 55%.

It should be pointed out that self-damping models are obtained on the basis of the best fit of measured data to a power law of the type reported in Table 1.1. Deviation of measured values from the fitted model averages around 20% in typical cases, and

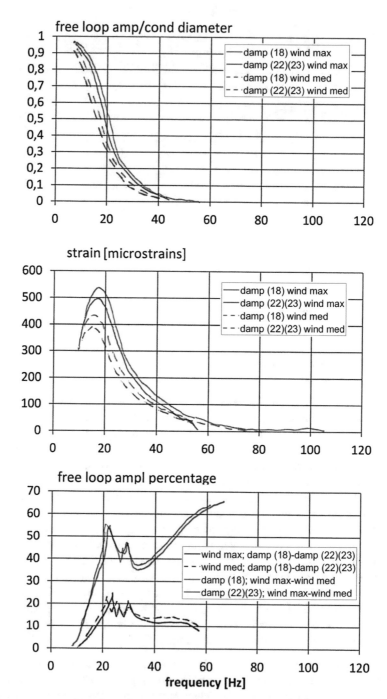

Fig. 1.13 Sensitivity of predicted amplitude (and strain) to uncertainties about wind power and self-damping (H/w = 1557 m, ACSR conductor, diameter = 23.55 mm)

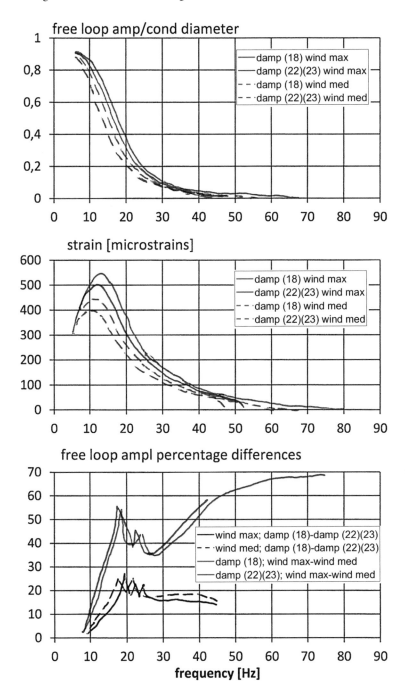

Fig. 1.14 Sensitivity of predicted amplitude (and strain) to uncertainties about wind power and self-damping (H/w = 1557 m, ACSR conductor, diameter = 28.11 mm)

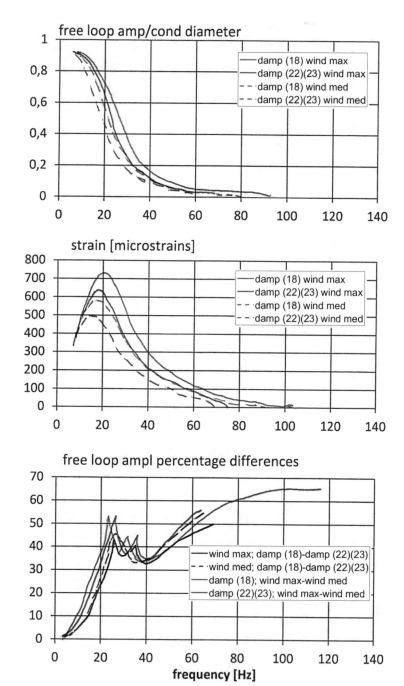

Fig. 1.15 Sensitivity of predicted amplitude (and strain) to uncertainties about wind power and self-damping (*H*/*w* = 2002 m, ACSR conductor, diameter = 23.55 mm)

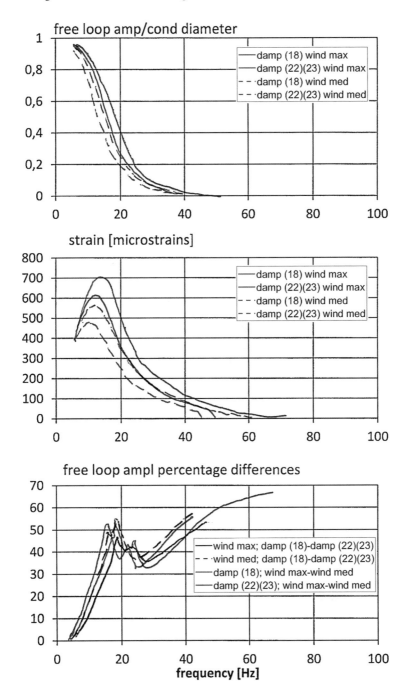

Fig. 1.16 Sensitivity of predicted amplitude (and strain) to uncertainties about wind power and self-damping (*H/w* = 2002 m, ACSR conductor, diameter = 35.10 mm)

is better in some ranges of frequency and worse in others, because of the difficulty of such measurements.

Obviously, the indicated sensitivity to the choice of wind power function and self-damping model would have been different had other such functions and models been included in the study. Other power functions are shown in Fig. 1.1. Greater variability due to the self-damping model would have been indicated had the Claren and Diana (1969) model been included.

1.6 Conclusions

This chapter has shown that through predictions of aeolian vibration level in operating lines, obtained using the EBP and variously available databases, it is possible to obtain a good reproduction of the frequency range and of the distribution of vibration amplitudes with frequency.

The predicted amplitude level in some cases is quite close to the measured, but in some other cases is quite different, due to several reasons. As already pointed out in the introduction, the EBP predicted amplitudes are the maximum amplitudes which can be excited by the wind on a certain conductor: if the mean wind speed (independently of the turbulence index) changes in time, the steady-state amplitude level can never be reached because the system is always in a transient condition. So the measured values are lower than the EBP predicted ones. Moreover, the measured frequencies may not perfectly agree with the Strouhal formula evaluated with the wind actual velocity, being related to a previous wind velocity. These facts could be responsible for some disagreement between vibration amplitude and frequency predicted and measured values when the wind structure is such that constant mean speed seldom occurs.

If the wind power functions and self-damping models employed in the study are indicative of the range of uncertainty in these parameters, then the range of uncertainty in EBP predictions of vibration amplitude is about ±50–60%.

Generally, the highest vibration strains are found at relatively low frequencies, where the contribution of dampers to overall damping is much more important than that of conductor self-damping. This is not true at high frequencies. Thus, the largest deviations between models for self-damping, which occur at the lowest frequencies are less important than they may appear.

In practice, assessment of the aeolian vibration condition of particular lines, with conductors whose mechanical properties are poorly defined, or with special terrain conditions, may require field measurements. Techniques to perform such measurements have been described previously (CIGRE SC22 WG11 TF2 1995).

Chapter 2
Modelling of Aeolian Vibrations of a Single Conductor Plus Damper

G. Diana, L. Cloutier, P. W. Dulhunty, M. Ervik, P. Hagedorn, C. Hardy, G. Kern, H-J. Krispin, A. Laneville, A. Leblond, A. Manenti, K. O. Papailiou, C. B. Rawlins and T. O. Seppa

2.1 Introduction

The first part of the TF research covered the modelling of aeolian vibrations of a single, undamped, conductor to define the vibration level and to assess the uncertainties connected to the present technology based on the energy balance principle.

It is well known that if the conductor tension (or, more precisely, the ratio between tension and conductor unit weight H/w) exceeds certain limit values, aeolian vibrations may cause serious conductor and fitting damage (CIGRE TF B2.11.04 2005). This limit H/w value is generally exceeded on transmission lines and then it is established a practice to protect conductors with suitable dampers.

For new transmission line designs, it is important to know how much additional damping is needed to control aeolian vibration within safe levels. To this purpose, various researchers have developed computation methods, based on the energy balance principle (EBP), to predict the aeolian vibration level of a conductor plus damper and then to allow for the selection of the suitable damping: these methods use the damper dynamic characteristics as measured on a shaker.

This chapter evaluates the computation methods, through direct comparison among them and with results obtained on an experimental span, with the final aim of defining the uncertainty of the considered methodology.

G. Diana (✉)
Department of Mechanical Engineering, Politecnico di Milano, Milan, Italy
e-mail: giorgio.diana@polimi.it

© Springer International Publishing AG 2018
G. Diana (ed.), *Modelling of Vibrations of Overhead Line Conductors*,
CIGRE Green Books, https://doi.org/10.1007/978-3-319-72808-7_2

2.2 Outline of the Methodology

Elements of the technology involved in the behaviour of damped single conductors are diagrammed in Fig. 2.1. The analytical elements are shown by the ellipses. Physical data entering into the analytical processes are shown by the rectangles on the left. Informational outputs from the analytical elements are shown by rounded rectangles. Sources of independent data to check or substitute for those outputs are shown by the rectangles on the right.

Each step in the analytical chain is affected by errors caused by assumptions and approximations required by the analytical procedures, and by inaccuracies in input

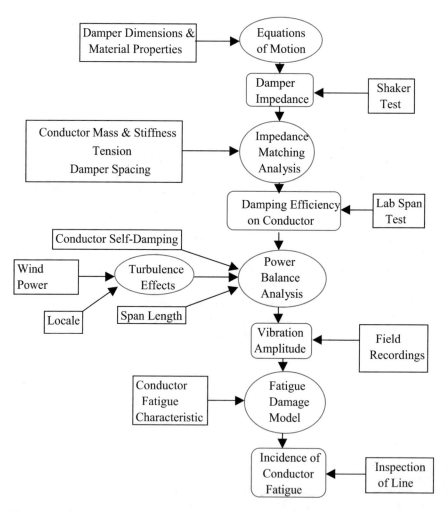

Fig. 2.1 Outline of the technology

data. Thus, accuracy deteriorates with progress down the chain. However, the accuracy of the final predictions can be improved by entering the chain with independent data at a lower point. For example, vibration amplitudes are more accurately determined from field recordings than from power balance analysis based on laboratory span testing of dampers. Damping efficiency on the conductor is more accurately determined by direct measurement on a laboratory span than on the basis of impedance matching analysis.

The present study assessed one branch of the technology chain. That branch started with shaker test data on actual dampers and ended with the predicted vibration amplitudes. These predictions were then compared with field recordings from a span equipped with dampers of the same type as tested on the shaker.

2.3 Analytical Models Used in the Study

Analytical modelling of conductor vibration with dampers is based generally on the Energy Balance Principle, which consists in determining the amplitude of vibration that leads to a balance between the energy introduced by the wind and that dissipated by the conductor and the Stockbridge-type damper.

The energy introduced by the wind and that dissipated by the conductor has already been defined in the first chapter. The aim of the present work is to evaluate the technology covering the dynamic behaviour of a span with a damper mounted on a conductor. Even though almost every prediction method is based on the energy balance principle, there are several differences among the analytical procedures used, depending on the adopted models and on the approach to the problem that is used. These differences cause variations among the analytical results obtained through different programmes, as it will be described in detail in the benchmark results paragraph.

The modelling considered here separates naturally into two steps. One concerns the interaction between the damper and the conductor, and determines how much dissipation the damper provides when actually mounted on the conductor. The second step, where energy balance is applied, determines the conductor vibration amplitude where power from the wind comes into balance with dissipation both by the damper and by conductor self-damping. Note that the first step can be avoided by simply measuring dissipation by the damper when mounted on a laboratory span (IEEE 1993; IEC 1998), obviously taking into account the difference between the actual and laboratory span length. We refer to this procedure as the 'basic method'. It has the advantage that various errors and assumptions of the analytical approach are avoided. It has the disadvantage of the significantly greater cost of testing. For the sake of uniformity, all participants in the benchmark study executed both of the above steps, relying entirely on damper performance characteristics as measured on a shaker.

The present work will not cover aspects connected to damper design, which has already been studied (Claren and Diana 1969; Tompkins et al. 1956). Thus, the first step of analytical modelling is the dynamic characterization of the damper. This is usually done by testing the damper on a shaker imposing a harmonic translational motion to the damper clamp and measuring the corresponding force acting on the clamp. The dynamic behaviour of the damper is then described by the transfer function between clamp velocity and the force (impedance) or between the clamp displacement and force (dynamic stiffness). The damper impedance (and the dynamic stiffness) is, of course, a function of the frequency.

A more complete characterization of the damper takes into account also the rotational motion of the clamp and the corresponding torque. Thus, a damper could be described in a more realistic way by a transfer matrix (impedance matrix) (Diana et al. 2003; Leblond and Hardy 1999) between a vector-valued input variable involving both the clamp velocity and angular velocity and a vector-valued output variable consisting of force and torque.

The type of damper chosen for the present study is the Stockbridge-type damper because it is widely used in all the countries.

It is well known that Stockbridge dampers have a significant nonlinear behaviour as a function of the clamp amplitude of vibration, due to the variation in the relative sliding among the messenger conductor wires; this modifies their flexural stiffness and damping, and, therefore, the damper dynamic behaviour. In order to take into account these nonlinear effects, the damper impedance is defined at a number of clamp vibration velocities (e.g. a range of 4–200 mm/s) and the intermediate velocity values are obtained through interpolation.

The conductor is normally modelled either as a string or a beam. The string model is easier to deal with because the parameters describing its behaviour (tensile load and mass per unit length) are easily identified, while the flexural stiffness of the beam model depends on the relative sliding among the wires forming the conductor. The generally adopted flexural stiffness value ranges from 30 to 50% of the conductor maximum stiffness value (i.e. the stiffness computed considering zero compliance at the wire contact interfaces) (Claren and Diana 1967). This flexural stiffness value is assumed constant with frequency and along the span.

Evaluation of the conductor self-damping has already been investigated in the first chapter; see also Table 2.1 where other references to the method adopted by each author are reported.

After defining the dynamical behaviour of the damper and of the conductor separately, the problems arising from their interaction have to be investigated. The major effect due to the presence of a damper on a conductor is the local distortion of the deflection shape of the conductor. The mathematical model of the system damper plus conductor must reproduce this distortion, in order to correctly evaluate the damper amplitude of vibration.

This is important because an error in this computation leads to an error of the transmitted force and of the energy dissipated. The damper nonlinear behaviour makes this aspect more critical, since the damper transfer function depends directly on the amplitude of vibration.

It is therefore important to reproduce as accurately as possible the local distortion of the conductor deflection shape. Moreover, if there is significant distortion,

Table 2.1 Benchmark comparisons for one damper on single conductor span

	Model 1	Model 2	Model 3	Model 4	Model 5
Reference	Diana et al.	Krispin	Leblond and Hardy	Rawlins	Sauter and Hagedorn
Configuration	Bersfort ACSR in 450 m span at 36 kN tension, damper at 1.7 m				
Damper characteristics	Figures 2.2, 2.3 and 2.4	Figure 2.2	Figure 2.2	Figures 2.2, 2.3 and 2.4	Figures 2.2, 2.3 and 2.4
Wind power data	Chapter 1	Rawlins (1983c)	Noiseux et al. (1986)	Rawlins (1983c)	Rawlins (1983c)
Consider turbulence	Diana et al. (1979)	Rawlins (1983a)	Noiseux et al. (1986)	Rawlins (1983a)	Yes
Self-damping data	Chapter 1	Chapter 1	Noiseux (1992)	Fitted to Chap. 1 data	Fitted to Chap. 1 data
Calculation method	Falco et al. (1973)	Hagedorn (1982)	Leblond and Hardy (2001)	Tompkins et al. (1956)	Hagedorn and Schmidt (1985)
Consider flexural stiffness	Yes	Yes	Yes/No	No	No
Consider damping rocking	Yes/No	No	Yes/No	No	No
Energy balance domain	Single freq.	Single freq.	Single central freq.	Single freq.	Single freq.
Model description	Complex modes	Travelling waves	Narrowband random travelling waves	Travelling waves	Travelling waves

due to a high level of force and torque transmitted by the damper, the local flexural stiffness of the conductor can undergo significant variations (Papailiou 1997; Diana et al. 2003) and this affects the damper behaviour and finally, the energy dissipated.

Although some researchers have developed calculation methods to account for all the terms of the damper impedance matrix (Diana et al. 2003; Leblond and Hardy 1999), the computation methods considered in the benchmark take into account only the force due to translation term of the damper impedance matrix. Two fundamental consequences result: on the one hand, the transmitted torque is completely neglected together with the force due to the rotation, and this leads to errors in the computation of the conductor deflection shape.

As already observed, this error can be avoided by measuring the energy dissipated by the damper on a laboratory span using the available measuring techniques [the Power Method, the ISWR method and the DEAM method developed by Rawlins (1988, 1997)].

Table 2.1 reflects the main differences among the methods adopted by the participants in the benchmark study: therefore, its results should be analysed taking into account this aspect along with the other differences.

2.4 Benchmark Results

The TF tested the existing technology by comparing amplitudes predicted by analysis against amplitudes actually measured in a field test.

Damper characteristics were obtained from direct measurements on a laboratory shaker. These were processed by five of the organisations represented on the Task Force, using their representations of the technology governing the interaction of the damper characteristics with the conductor, and their selected functions for aeolian and self-damping power. The results of the calculations were compared with field recordings from a span fitted with a damper of the same design as used in the laboratory shaker test.

Three commercially available dampers of the same design as the one installed on the field span, but not including the sample mounted on the field span, were tested on the shaker at different amplitudes of clamp velocity. The measured dynamic stiffness characteristics are reported in Figs. 2.2, 2.3 and 2.4 for damper A, B and C, respectively. Each figure reports a number of curves corresponding to the different tested velocities. A great variation of the transfer functions, both in terms of resonant frequency shift and corresponding amplification factors, is observed. If the moduli of these characteristics for the three damper samples, measured at the same velocity (e.g. 10 mm/s) are compared, the observed dispersion is quite significant. This can make an important contribution to the uncertainty of the computations with respect to the experimental values. Data describing the measured field span are reported in Table 2.2.

The measured vibration amplitudes are reported in the lowest curve of Fig. 2.5. Predicted amplitudes without a damper, computed in Chap. 1, are shown in Fig. 2.5 as well, for two turbulence levels (5 and 15%).

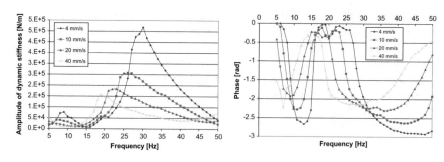

Fig. 2.2 Dynamic stiffness of the damper A

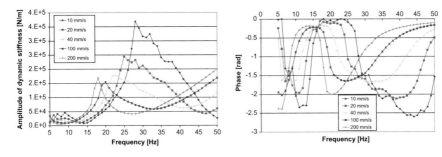

Fig. 2.3 Dynamic stiffness of the damper B

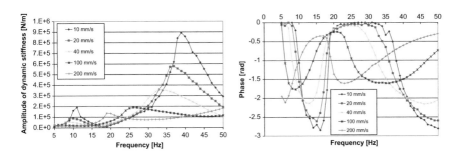

Fig. 2.4 Dynamic stiffness of the damper C

Table 2.2 Main data of the measured span (Leblond and Hardy 1998)

Conductor	Type	ACSR Bersfort (48/7)
	Diameter	35.6 mm
	Mass per unit length	2.37 kg/m
	Tension	36 kN
Span length	450 m (suspension)	
Type of terrain	Open, flat, no obstruction, with snow cover (farmland)	
Position of the damper	One damper per span located 1.7 m from centre of the suspension clamp	

The participants in the benchmark study used the methods and assumptions listed in Table 2.1, to predict the expected amplitudes in the field span. For this purpose, they employed the combinations of variables listed in Table 2.3.

The results of these calculations are presented in Figs. 2.6, 2.7, 2.8, 2.9, 2.10, 2.11 and 2.12.

It is clear from an examination of these figures that none of the predictions agreed well with the recorded field data over the entire frequency range. Furthermore, the quality of agreement is sensitive to the combination of variables

Fig. 2.5 Measured amplitude
of vibration and analytical
computation of the amplitude
without damper with two
turbulence levels

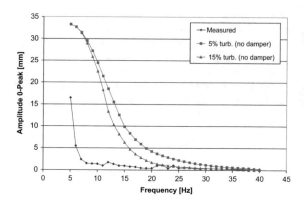

Table 2.3 Parameter values covered

	Model 1	Model 2	Model 3	Model 4	Model 5
Damper sample	A, B, C	A	A	A, B, C	A, B, C
Turbulence (%)	5, 15	10, 15	13	5, 10, 15	1, 5, 15
Conductor stiffness[a]	0.05, 0.5	0.015, 0.5	0.05	0	0

[a]Fraction of EJ_{max}, the flexural stiffness computed with no interstrand slipping

Fig. 2.6 Model 2 results

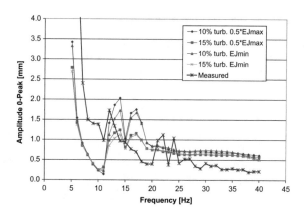

considered in each case and, perhaps, varies among the participants because of
differences noted in Table 2.1.

It is difficult to assess the relative importance of these factors (combination of
variables and differences in method and in assumptions) without some numerical
measure of the quality of agreement between predicted and measured amplitudes.
To provide such a measure, we have computed, for each frequency, the error factor
between analytical computation and experimental results:

Fig. 2.7 Model 5 results

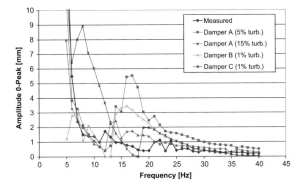

Fig. 2.8 Model 3 results

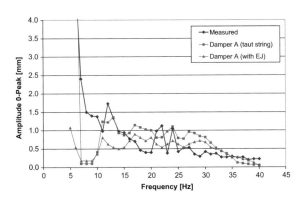

Fig. 2.9 Model 4 results
(damper A)

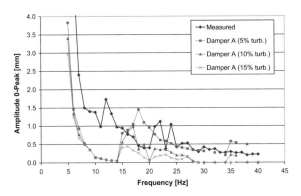

$$\text{err}_a(i) = \frac{A(i) - M(i)}{M(i)}; \ i - 1 \ldots N, \tag{2.1}$$

where i is the counter of each frequency, N is the total number of frequencies, $A(i)$ is the vibration amplitude computed with respect to the ith frequency and $M(i)$ is the experimental one.

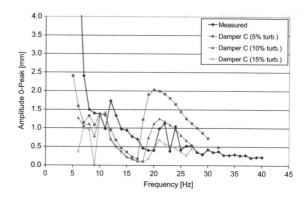

Fig. 2.10 Model 4 results (damper C)

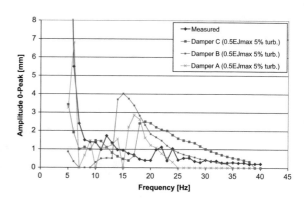

Fig. 2.11 Model 1 results ($0.5EJ_{max}$—T 5%)

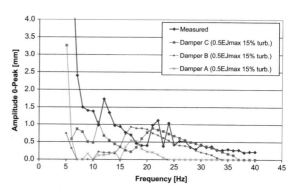

Fig. 2.12 Model 1 results ($0.5EJ_{max}$—T 15%)

In order to compare the error with the amplitude level leading to conductor fatigue, another error factor has been defined according to the following relation:

$$\text{err}_b(i) = \frac{A(i) - M(i)}{\text{Ref}(i)}; \; i - 1 \ldots N, \tag{2.2}$$

where Ref(i) has been chosen equal to 33/f (mm).

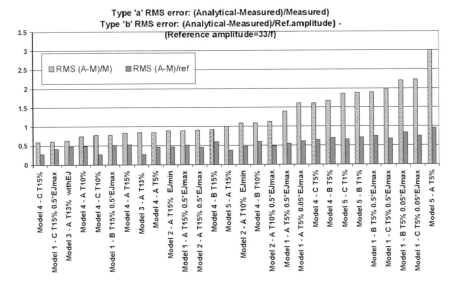

Fig. 2.13 RMS error

The computation of the root mean square (RMS) of the error factor (Fig. 2.13) then gives a useful indication of the quality of agreement between the analytical predictions and the vibration actually measured in the field span:

$$\text{RMS}\left(\text{err}_{a,b}\right) = \sqrt{\frac{1}{N} \sum_{i=1...N} \text{err}_{a,b}(i)^2} \qquad (2.3)$$

Computing the RMS value for every calculation performed by each participant, as reported in Fig. 2.13, shows that the dispersion of the quality of results is significant. Part of this variation is due to the combinations of variables shown in Table 2.3. It can be observed that all the computations performed with a turbulence level higher than 10% show an RMS error (type a) below 1.1, while the computations with the turbulence of the order of 1–5% show an RMS error (type a) above 1.4.

On the other hand, differences in analytical method and assumptions also contribute to the dispersion in quality of the predictions, although the impact seems to be smaller than for the effect of differences among turbulence level and damper samples A, B and C. Figure 2.14 shows, for example a comparison between the two computations with the lowest RMS (type a) values in Fig. 2.13, both performed under the same conditions (damper C and 15% of turbulence level). One can see that the analytical curves can be nearly superimposed and they are close to the experimental one. Figure 2.15, on the other hand, shows a comparison of all calculations regarding damper A with high turbulence level and assumed conductor flexural stiffness. The dispersion of the curves is clearly higher than the previous

Fig. 2.14 First comparison: Damper C with T 15%

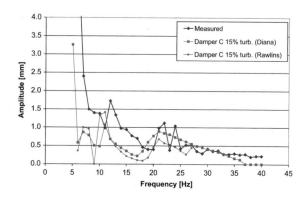

Fig. 2.15 Second comparison: Damper A with T 15%

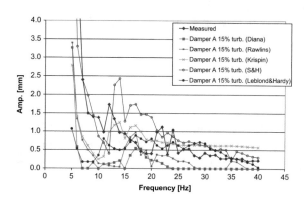

example, but is much less than for the effect of turbulence level and sample choice among A, B and C.

Figures 2.16 and 2.17 show the calculated vibration strains at the suspension clamp for the cases represented in Figs. 2.14 and 2.15. Also included as the upper two curves are the calculated strains corresponding to the undamped cases of Fig. 2.5. These strains are those on the conductor at the suspension clamp at the span extremity not fitted with damper.

Figures 2.16 and 2.17 show that, even if the strains predicted by the different participants exhibit a certain variability, the computed strain level reduction due to the damper presence (with respect to the undamped conductor), in any case, is of the same order of magnitude as the one defined by the field measurements.

Note, however, that in attempting to apply analysis in the design of new lines, the important comparison is between the predicted strain levels and the levels that can cause fatigue, rather than merely the undamped levels. If the design makes full use of the damper capabilities, then expected strains will approach maximum allowable levels. The type b error factors shown in Fig. 2.13 provide a guide to the safety factors that must be applied to the maximum allowable levels.

Fig. 2.16 Strain computation for the first comparison

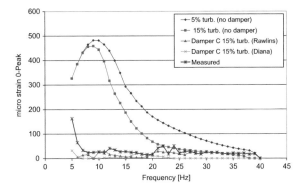

Fig. 2.17 Strain computation for the second comparison

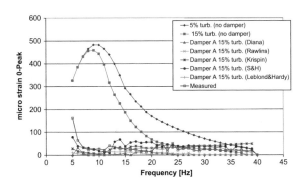

2.5   Discussion

Benchmark results show a wide dispersion of the predicted maximum amplitude values, and large discrepancies between the analytical predictions and the measured behaviour of the test span. These differences arise from several sources.

1. The different participants relied on different sets of assumptions (Table 2.1).
2. The steady-state or stationary condition assumed in the analytical approach could be a possible cause of the discrepancy between analytical predictions and the measured data since actual vibration is the result of a complex interaction between the conductor and the turbulent wind, usually with many frequencies simultaneously excited, rarely reaching a steady-state or stationary condition (Diana et al. 1993).
3. The boundary condition at the supporting structure in situ, is between a perfect hinge and a perfect fixed joint. As the damper is installed close to the span end, the boundary condition can modify the conductor deflection shape significantly, and, consequently, the energy dissipated by the damper.

Fig. 2.18 Antinode
amplitude computation from
field-measured damping
efficiency

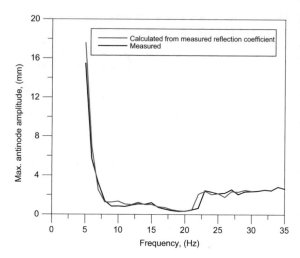

4. A major source of inaccuracy, in predicting expected amplitudes in the field, is
   the calculation of the interaction of the damper with the conductor. This calculation uses the damper characteristics, as measured on a shaker, to predict the
   power the damper will dissipate in situ on the conductor. All of the calculation
   methods referred to in Table 2.1 included this step in some form. Note that
   dissipation in situ can be, and often is, measured directly in laboratory spans
   (IEEE 1993; IEC 1998; Rawlins 1983b): this method reduces the
   above-mentioned uncertainties related to the conductor–damper interaction.

The DEAM method (Rawlins 1988) was applied for measurements on a field span
by Leblond and Hardy (1997), in order to determine damper plus extremity dissipation in situ. They calculated predicted antinode amplitude on the basis of the
EBP, using their own assumptions for the wind energy input. It was clearly shown
(Leblond and Hardy 2000) that using the field-measured damping efficiency in the
calculation model yields predicted antinode amplitudes which are impressively
close to those measured on the test line as shown in Fig. 2.18.

While the dispersion of the dynamic stiffness of different damper samples shown
in Figs. 2.2, 2.3 and 2.4 were reflected in corresponding dispersion in predicted
amplitudes as shown in Figs. 2.6, 2.7, 2.8, 2.9, 2.10, 2.11 and 2.12, it is not a
reflection of the technology. This dispersion will remain the same, independently of
the technology used.

2.6 Future Work

Several sources have been identified for the dispersion among calculated results and
the large discrepancies between analytical predictions and measured behaviour
revealed by the study. Further studies are needed for the purpose of developing

recommendations for reducing or eliminating these errors, thus permitting the design of more reliable vibration protection.

2.7 Conclusion

The strains predicted by the different researchers exhibit considerable variability. Nevertheless, analytical methods based on the EBP and shaker-based technology can provide a useful tool for use in the design of damping systems for the protection of single conductors against aeolian vibrations. It should be used with circumspection and be supplemented by references to field experience.

Chapter 3
Modelling of Aeolian Vibrations of Single Conductors Strung at Relatively High Tensile Load

G. Diana, H. J. Krispin, P. Bousseau, S. Gelderblom, P. Hagedorn, D. Hearnshaw, M. Landeira, A. Leblond, A. Manenti, C. B. Rawlins, J. B. Wareing, P. Catchpole, G. Chapman, U. Cosmai, O. Cournil, M. Ervik, T. A. Furtado, C. Hardy, T. Leskinen, M. Rivero, V. A. Shkapsov, S. Thaddey, J. A Araujo, J. M. Asselin, G. E. Braga, L. Cloutier, D. G. Havard, S. Kolosov, A. Laneville, P. Van Dyke and A. Vinogradov

3.1 Introduction

The first two chapters cover the modelling of aeolian vibration of an undamped single conductor and of a single conductor plus one damper. This chapter defines the expected vibration level and assesses the uncertainties associated with the present technology based on the energy balance principle (EBP) and a shaker-based approach. It has been shown that analytical methods based on the EBP and a shaker-based technology can provide a useful design tool for damping systems that protect a single conductor against aeolian vibration.

The purpose of the present chapter is to evaluate the effectiveness of these methods for the design and/or verification of the damping system of long, single conductor spans strung at relatively high tensile load, such as crossings, which need more than one damper per span extremity to be effectively damped against aeolian vibration.

As in the first two chapters, this chapter is based on an analysis of the available technology and on the results of two benchmarks: an analytical–analytical benchmark and analytical–experimental one. The comparison between the analytical results produced by the different available models and the experimental one will help to understand the limitations and the usefulness of the approach.

Some difficulty in finding the data for the analytical–experimental benchmark has been experienced by the WG members and consequently, the specific case of the

G. Diana (✉)
Department of Mechanical Engineering, Politecnico di Milano, Milan, Italy
e-mail: giorgio.diana@polimi.it

© Springer International Publishing AG 2018
G. Diana (ed.), *Modelling of Vibrations of Overhead Line Conductors*,
CIGRE Green Books, https://doi.org/10.1007/978-3-319-72808-7_3

Messina crossing in Italy was chosen but this does not limit the validity of the assessment.

The application of the EBP technology is more critical for long spans than for normal length spans discussed in the two previous chapters. The EBP in the previous chapters is based on a constant mean wind speed along the entire span but the longer the span length, the more unrealistic it would be to have a constant mean wind along the entire span, especially at the low wind speeds required to produce aeolian vibration. The tensile load variation along the span also becomes significant for large sags and it affects the conductor vibration wavelength.

It must be pointed out that EBP-based methods do not simulate the full complexity of the problem. More sophisticated methods (Rawlins 2000; Giorgio et al. 2009, 2010) can be used to account for the effect of the mean wind speed variation in time and space and wind turbulence, but these are still at a research stage and are not within the scope of this work.

However, the use of the EBP approach, i.e. the assumption of a constant wind speed along the span, for long span applications should guarantee predicted vibration amplitudes higher than those that occur in reality, therefore producing conservative damping system designs.

Outline of the technology

Elements of the technology involved in the behaviour of damped single conductors have been already presented in Chap. 2. Even if the present case relates to long spans with many dampers at the extremities, the same considerations reported in Chap. 2 hold.

Some long span effects have been taken into account in (Rawlins 2000), where it was shown that they reduce the required span-end damping in long spans to limit the severity of vibration to particular values, compared to requirements estimated from conventional technology ignoring those effects. One effect arises from the significant variation in vibration wave amplitudes that occurs over long spans; another effect is caused by nonlinearity in the self-excitation mechanism. For spans, such as a river or a fjord crossing, which are long enough to require span-end damping near the limits of what is feasible in practice, it was shown that the reduction in damping required may be as much as 25%. For these longer spans, in-span damping arrangement is necessary to adequately damp out the wind energy input. In-span damping is achieved by placing groups of dampers at intervals along the span, say at the one-third points, to effectively divide long spans into several shorter and more controllable subspans.

Some other effects may also lead to errors in the analysis of long spans subjected to high tensile loads and may require future research work to improve the EBP-based technology. These will be addressed in the 'Discussion'.

3.2 Analytical Methods and Assumptions Used in the Study

The study was carried out by comparing the results of EBP calculations by several different experts. However, there were nominal differences in analytical procedures. All experts based their calculations on the same data sets describing the characteristics of the dampers at the ends of the span, but there was some variation in the wind power input functions and self-damping models applied in the EBP calculations, as well as in the modelling of the effects of wind turbulence. Table 3.1 summarizes these similarities and differences. Although there are differences in the 'approach' they are all based on the same assumptions.

Table 3.1 Analytical methods and assumptions used in the study

	Benchmark	Diana et al.	Krispin	Leblond	Rawlins	Landeira
Damper characteristics	AA + AE	Figure 3.1	Figure 3.1	Figure 3.1	Figure 3.1	Figure 3.1
Wind power data		Chapter 1	Rawlins (1983c)	Rawlins (1983)	Rawlins (1983c)	Chapter 1
Turbulence		Diana et al. (1979)	Rawlins (1983a)	Noiseux et al. (1988)	Rawlins (1983a)	Diana et al. (1979)
Self-damping data	AA	Chapter 1	Chapter 1	Noiseux (1992)	Estimate	Chapter 1
	AE	Table 3.3	Table 3.3	Table 3.3	Table 3.3	–
Calculation method		Falco et al. (1973)	Hadulla (2000)	Leblond and Hardy (2001)	Tompkins et al. (1956)	Falco et al. (1973)
Approach		Transfer complex constants	Transfer impedance Complex eigenmodes	State vector and transfer matrix method	Transfer impedance	Transfer complex constants
Flexural stiffness		$0.5\ EJ_{max}$	$0.5\ EJ_{max}$	$0.5\ EJ_{max}$	No	$0.5\ EJ_{max}$
Damper rocking		No	No	No	No	No
Damper nonlinearity		No	No	No	No	No
Long span effect		No	No	No	Rawlins (2000)	No
Energy balance domain		Single freq.	Single freq.	Single central freq.	Single freq.	Single freq.
Mode description		Complex modes	Travelling waves Complex modes	Narrowband random travelling waves	Travelling waves	Complex modes

Notes: *AA* stands for analytical–analytical benchmark, while *AE* stands for analytical–experimental benchmark

Damper rocking and damper nonlinearity were not considered by any participant because no data on these effects were available for the dampers used in the study

3.3 Benchmark Results

The Working Group tested the existing technology by two different benchmarks: the first one is an analytical–analytical benchmark, i.e. the amplitudes predicted by the various available models applied to a certain test case are compared one against the other; the second one is an analytical–experimental benchmark: in this case amplitudes predicted by analysis are compared to amplitudes actually measured in a field test. The experimental data were available to the benchmark participants since the beginning.

The same two damper types have been used for the two benchmarks: their dynamic stiffness is reported in Fig. 3.1. The damper stiffness has been measured on a shaker at 100 mm/s clamp vibration peak velocity. Measurements at different amplitudes of clamp velocity were not available and so the damper nonlinearity could not be taken into account in evaluating the conductor plus damper response. Data concerning the damper production variability were not available (Note: Tests made on different samples of the same damper type produce a dispersion of data peculiar of the production process, which can be statistically analyzed), nor was the moment transmitted by the damper clamp.

3.4 Analytical–Analytical Benchmark

Data describing the considered trial case are reported in Table 3.2.

For this analytical–analytical benchmark each model uses its selected functions for wind, conductor self-damping power and conductor bending stiffness as shown in Table 3.1.

The predicted vibration amplitudes and the predicted system damping are reported in Fig. 3.2a, b.

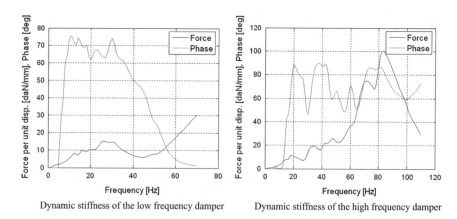

Dynamic stiffness of the low frequency damper   Dynamic stiffness of the high frequency damper

Fig. 3.1 Damper dynamic stiffness modulus and phase

Table 3.2 Analytical–analytical benchmark input data

Conductor	AACSR (91 × 2.35 + 78 × 2.35) diameter: 35.25 mm, mass: 3.935 kg/m
Tension	185 kN; catenary constant: 4792 m
Wind	No or very low turbulence
Span length	2000 m; suspension–suspension
Dampers	three dampers at both extremities, symmetrically placed first low-frequency damper at 1.80 m from the suspension clamp second high-frequency damper at 0.90 m from the first damper third high-frequency damper at 1.50 m from the second damper

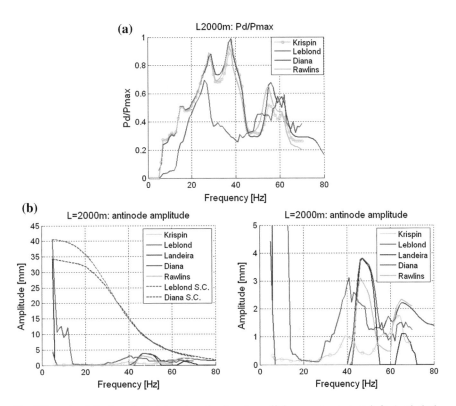

Fig. 3.2 a Analytical–analytical benchmark: damping efficiency per span end. **b** Analytical–analytical benchmark: predicted amplitudes for the undamped and damped conductor

The participants in the benchmark study used the methods and assumptions listed in Table 3.1 to predict the expected amplitudes in the trial case.

From the examination of these figures, the results appear in reasonable agreement. The variability of the predicted amplitude in terms of 3 times the standard deviation, at least for the models in better agreement, has a maximum value of approximately 1.5 mm, compared with a mean amplitude of 3.5 mm. This could be interpreted, that there is little difference among these models.

The significance of this benchmark, being only an analytical–analytical benchmark, is not compromised by the fact that the predicted amplitudes account for neither the damper nonlinearity nor the damper production typical variability and for the effect of the rotation of the damper clamp, because the same damper data are used by all the models.

It is also important to note that at least three of the four compared models give very close damping efficiencies (Fig. 3.2b), thus demonstrating that for the three dampers positioned on the conductor, those models reproduces the same contribution to the total damping.

Analytical–experimental benchmark

Data describing the considered field span are listed in Table 3.3 (Note: The field span is that of the Messina crossing in Italy, which was replaced by a 400 kV submarine conductor in the 80s).

Table 3.3 Analytical–experimental benchmark: input data

Conductor Tensile load Span length	Diameter: 26.80 mm, steel area: 305 mm^2; aluminium area: 45 mm^2, mass: 2.70 kg/m, UTS: 523 kN 220 kN (maximum, constant value, tensile load controlled by counterweights). Catenary constant: 8306 m 3646 m, suspension–suspension (conductor on pulleys—see Fig. 3.6)
Wind turbulence	Turbulence index $I_t \approx 0.1$ for wind speeds in the range 5–10 m/s (Falco et al. 1973)
Dampers	six dampers at both extremities, symmetrically disposed: first high-frequency damper at 2.00 m from the suspension clamp second high-frequency damper at 0.75 m from the first damper third high-frequency damper at 1.75 m from the second damper fourth low-frequency damper at 2.50 m from the third damper fifth high-frequency damper at 1.00 m from the fourth damper sixth low-frequency damper at 1.50 m from the fifth damper
Accelerometers	A1 at 2.00 m from the sixth damper into the span A2 at 2.00 m from A1 A3 at 1.00 m from A2
Conductor self-damping	Energy dissipated by the conductor per unit length of conductor (Falco et al. 1973) $E = 2 \times 10^{-3} \, f^{3.5} \, u^{2.4}$ (J/m) f: vibration frequency (Hz), u: antinode vibration amplitude m 0-peak

Six dampers per span extremity are considered, with symmetrical disposition, as reported in Table 3.3. The two damper types installed in the crossing are the same used for the analytical–analytical benchmark.

For the analytical–experimental benchmark, each model uses its selected function for wind power input but all the models use the same function for the conductor self-damping: see Tables 3.1 and 3.3 (5th row).

Experimental results for this trial case are available in the form of antinode amplitude (0-peak) of vibration, deduced from the measurements of the three accelerometers A1, A2, A3 positioned as indicated in Table 3.3 (4th row). Experimental results are reported in Fig. 3.4, together with the predicted amplitudes. For a comparison with the output of an EBP-based model, experimental results are only reliable up to about 40 Hz. As a matter of fact, the measured wind speed related to frequencies higher than 40 Hz was not constant along the span and in this case, as pointed out also in the introduction, the measured conductor response is not comparable to an EBP prediction.

The benchmark results are reported in Figs. 3.3, 3.4 and 3.5.

A sensitivity analysis has been performed to try to better understand the origin of the discrepancies between analytical and experimental results. Two main factors have been considered: the first is the error in the dampers' position in terms of distance of the first damper from the suspension clamp (due to the fact that, as reported in Table 3.3, first row, the conductor is on pulleys—Fig. 3.6) and the second is the effect of the damper nonlinearity. While the first factor was shown to have a negligible influence on the system response, the change of the damper dynamic stiffness due to the damper nonlinearity significantly affects the system response.

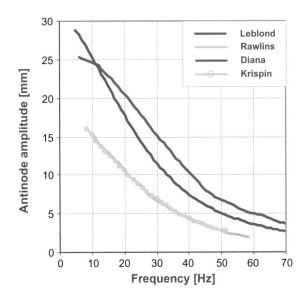

Fig. 3.3 Analytical–experimental benchmark: antinode amplitude, undamped conductor

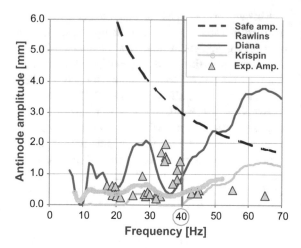

Fig. 3.4 Analytical–experimental benchmark: antinode amplitude, damped conductor

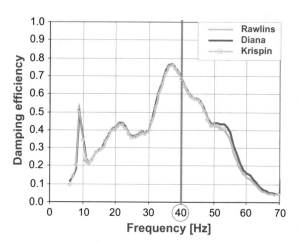

Fig. 3.5 Analytical–experimental benchmark: damping efficiency per end

Analyzing Figs. 3.3, 3.4 and 3.5, it can be observed that the damping efficiency (Fig. 3.5) is the same for the three models, even though they are based on the different approaches (see Table 3.1) thus confirming the observations for the analytical–analytical benchmark (Fig. 3.2b).

It can also be observed that, as the conductor self-damping is common to all the simulations, the differences in the predicted amplitudes arise mainly from the different formulations of the wind power input. This is clearly evident for the undamped conductor in Fig. 3.3. The higher wind power input produces higher predicted amplitudes for the damped conductor, as shown in Fig. 3.4.

The experimental data seem to suggest that, at least in this case, the real wind power input should be in between the Diana assumption (Chap. 1 and Diana et al. 1979) and the Rawlins/Krispin assumption (Rawlins 1983c).

Fig. 3.6 Analytical–
experimental benchmark: the
suspension set of the Messina
crossing

3.5 Discussion

As with the two previous chapters, this chapter is an assessment of the technology
used in modelling aeolian vibration of overhead conductor spans. It goes beyond in
that it tests that technology in the more challenging engineering environment of
very long spans strung at relatively high tensions. That environment is marked
mainly by three differences from those in the first two chapters.

- First, the length and sag range of long spans are such that the wind field should
  not be considered homogeneous also taking into account that as reported in
  (Giorgio et al. 2010), the mean wind speed is changing in time. Gust widths are
  much shorter than span lengths, and the vertical gradient in wind speed can
  expose the middle of the span to significantly lower wind velocities than are
  found simultaneously near the ends. As pointed out in the introduction, methods
  to take into account the non-homogeneity of the wind field are still at best in the
  research stage (Giorgio et al. 2009, 2010), therefore, none of the participating
  experts considered its effects. Similarly, effects of variation in tension along the
  span have not been taken into account. Effects of turbulence (homogeneous
  turbulence in the wind) were taken into account, as noted in Table 3.1.

- Second, again due to the big sag, tension at the support can be significantly greater than at midspan. This can cause the coupling between the vortex induced forces and the mechanical system (conductor), as well as self-damping, to vary along the span.
- Thirdly, calculation of damping efficiency of multi-damper arrangements at the ends of the span is more complicated than with only a single damper.

Calculation of damping efficiency of multi-damper setups is more complicated than for single damper installations because of the interaction of the dampers. However, Figs. 3.2b and 3.5 show generally quite good agreement among the participants. This indicates that these calculations using shaker-based data on damper characteristics rest on sound technology. This is a significant finding of the present study.

On the other hand, those calculations also clarify the need for data on damper characteristics as a function of the damper clamp amplitude. The details of the calculation results revealed large differences in predicted clamp amplitudes among the dampers under consideration during steady vibration at a constant frequency, for most of the frequencies covered. Furthermore, these predicted clamp amplitudes were in general very different from the amplitudes on which the damper characteristics were based and, as already observed, represent one of the reasons for the discrepancies between experimental and analytical results. This differs from the study in Chap. 2, where damper characteristics were provided as a function of clamp amplitude.

3.6 Conclusions and Future Work

Future research work is needed to improve the EBP technology, which generally produces a safe design of the damping system. In Fig. 3.4, for instance, the amplitudes of vibration expected on the base of the damping system designed through the EBP are reported together with the measured amplitudes and the safe amplitudes, evaluated through the generally accepted limit of $fy_{max} = 118$ mm/s (EPRI 2009). The measured amplitudes are lower than the safe amplitudes.

Future work is needed in order to better understand the effect of turbulence and mean wind speed variation, together with a better simulation of the mechanical system, in such a way to reproduce tensile load variations and multi-frequency excitation.

Chapter 4
Modelling of Aeolian Vibrations of Bundled Conductors

G. Diana, H. J. Krispin, J. Havel, J. Huang, J. Garnier,
S. Kolosov, J. P. Paradis, A. Leblond, C. B. Rawlins, U. Cosmai,
P. Van Dyke, J. L. Lilien, W. Troppauer, A. Manenti, L. Mazzola,
E. Ruggeri, N. Sahlani, D. Proctor, C. Rozé, M. Landeira, S. Thaddey,
C. Hardy, R. Tong, B. Liu, A. Bhangor, A. Vinogradov,
T. A. Furtado and M. Araujo

4.1 Introduction

One of the purposes of this chapter is to evaluate the effectiveness of analytical methods based on the energy balance principle (EBP) and a shaker-based technology for the design and/or verification of the damping system of conductor bundle spans with respect to aeolian vibrations.

As in the case of the previous chapters, this chapter is based on an analysis of the available technology and on the results of benchmarks: an analytical–analytical benchmark and an analytical–experimental one are used for the evaluation relevant to aeolian vibrations. The comparison between the analytical results produced by the different available models and the experimental one will help to understand the limitations and the usefulness of the considered approach.

Field tests relevant to a 500 m quad bundle span equipped with ACSR Drake conductor and spacer dampers have been selected as test case both for the analytical–analytical and analytical–experimental benchmark for the aeolian vibrations evaluation. The selected case has an H/w parameter around 2000 m which is below the 2500 m recommended safe design tension (CIGRE TF B2.11.04 2005).

It must be pointed out that EBP-based methods do not simulate the full complexity of the problem. The complexity of the problem and the limitations of the EBP approach, together with the more sophisticated tools nowadays available to reproduce the aeolian vibrations phenomenon are reported in the first three chapters.

However, from the engineering point of view, the EBP is a very useful tool for the design of damping system necessary to control aeolian vibrations.

G. Diana (✉)
Department of Mechanical Engineering, Politecnico di Milano, Milan, Italy
e-mail: giorgio.diana@polimi.it

© Springer International Publishing AG 2018
G. Diana (ed.), *Modelling of Vibrations of Overhead Line Conductors*,
CIGRE Green Books, https://doi.org/10.1007/978-3-319-72808-7_4

4.2 Analytical Methods and Assumptions Used in the Study

The study was carried out by comparing the results of EBP calculations coming from three different experts. All experts based their calculations on the same data sets describing the characteristics of the spacer dampers, however, there were nominal differences in analytical procedures, such as the wind power input functions, self-damping models applied in the EBP calculations, or the modelling of the effects of wind turbulence. Table 4.1 summarises these similarities and differences taken as main assumptions in the simulations. Although the above differences, all the models are based upon the same assumptions and they are developed in the frequency domain, hence, a preliminary evaluation of the bundle natural frequencies and modes of vibration is necessary as shown in Chaps. 1 and 2. To do this, for each mode of vibration the aeolian vibration amplitudes are achieved through a balance between the power input from the wind, the power dissipated by the conductors, and the damping devices (spacers and dampers).

In all the models the following assumptions are considered:

- The vertical and horizontal components of the conductor motion are taken into account. The wind power input is only related to amplitudes in the vertical plane (Brika and Laneville 1997);
- The conductor self-damping is due to amplitudes resulting from vertical and horizontal motion;
- The spacer dampers are modelled through their inertial, elastic and damping characteristics, as detailed in Table 4.3. The coupling among subconductors is defined by the spacer dynamic response, which is a function of the vibration frequency;
- The aerodynamic damping related to the horizontal motion is not considered;
- The torsional stiffness of the conductor is not considered in the computation of the bundle natural frequencies;
- The longitudinal motion is not considered;
- The energy input from the wind comes from wind tunnel tests on two vibrating cylinders, with one in the wake of the other (Diana et al. 1976; Brika and Laneville 1997; Belloli et al. 2003):

  - Half of the overall energy derived from the wind tunnel tests on two cylinders is applied to any conductor of the bundle (independently on the subconductors number)
  - Because the energy is a function of the amplitude of motion of the conductor, each subconductor may have different wind power input

In the following, a brief description of the three models considered is provided.

Table 4.1 Analytical methods and assumptions used in the study (numerical table entries refer to supporting references)

	Benchmark	Analytical methods		
		Diana et al.	*Krispin*	*Cosmai (Claren)*
Case data	AA + AE	Table 4.2	Table 4.2	Table 4.2
Spacer characteristics	AA + AE	Table 4.3	Table 4.3	Table 4.3
Wind power data	AA	Curve b Table 4.4 (Diana et al. 1982)	Curve b Table 4.4 (Diana et al. 1982)	
	AE	Table 4.4 (Belloli et al. 2003; Diana et al. 1982)	(Anderson and Hagedorn 1995; Hadulla 2000)	Curve b Table 4.4 (Diana et al. 1982)
Consider turbulence		YES (Belloli et al. 2003; Diana et al. 1982)	NO	YES (Variable turbulence)
Self-damping data	AA	Table 4.4	Table 4.4	
	AE	(Claren et al. 1971)	(Anderson and Hagedorn 1995; Hadulla 2000)	(Claren et al. 1971)
Calculation method		(Claren et al. 1971)	(Anderson and Hagedorn 1995; Hadulla 2000)	(Claren et al. 1971)
Approach		Matrix transfer method. Modal approach	Complex Eigenmodes	Matrix transfer method Modal approach
Consider flexural stiffness		0.5 EJ_{max}	0.5 EJ_{max}	0.5 EJ_{max}
Tensile load differentials	AA	No	No	
Tensile load differentials	AE	Yes	Yes	No
Energy balance domain		Single freq.	Single freq.	Single freq.
Spacer model		(Claren et al. 1971)	(Anderson and Hagedorn 1995; Foata and Noiseux 1991)	(Claren et al. 1971)
Possible to consider dampers		Yes	Yes	Yes
Possible to consider armour rods		Yes	No	Yes

Note: *AA* stands for analytical–analytical benchmark, while *AE* stands for analytical–experimental benchmark

Krispin Model

The computational model determines vibration modes of the conductor bundle (Anderson and Hagedorn 1995). For that purpose, the subconductors are assumed to behave like strings with small bending stiffness. The subconductors are divided into subspans at the locations, where spacers are attached. The spacers are represented by an impedance matrix which describes the relation between the conductor forces and velocity at the spacer clamps. Stockbridge dampers are treated analogously.

Formulating the equation of motions of the subspans and taking into account boundary conditions (clamped span ends) and compatibility conditions at the spacer clamps, leads to a set of homogeneous equations. Solving the eigenvalue problem gives the complex eigenvalues. The corresponding eigenvectors define the complex eigenmodes.

The EBP method is employed to evaluate the vibration amplitude of these modes. Complex mode shapes have antinode and node amplitudes that vary along the span. The complex mode shape is approximated by an equivalent standing wave possessing the same mechanical energy as the complex wave. Conductor self-damping and wind power input are evaluated for the amplitudes of this equivalent standing wave. Power dissipation of the spacers is derived from the clamp velocities and the spacer impedance.

Vibration intensity is calculated in terms of the maximum value of fy_{max}, which is the product of frequency and antinode amplitude in the respective subspan. Nominal strains at clamping points are derived by using well-known relationships between nominal strain and fy_{max}.

Diana Model

In this model, eigenfrequencies and eigenmodes of the system (real modes) are computed through the matrix transfer method. Field matrices $[B_i]$ ($i = 1,..., N$ with N = number of subspans) defining the relationship between displacements and forces on the conductors at the extremities of each subspan and point matrices $[P_j]$ ($j = 1, ..., N - 1$) defining the relationship between displacement and forces on the conductors right side and left side of each spacer are defined. The product of all the field and point matrices gives a matrix $[A]$ defining the relationship between displacements and forces at the span extremities. End conditions allow for the computation of the eigenfrequencies and correspondent eigenmodes (Belloli et al. 2003; Diana et al. 1979, 1982; Claren et al. 1971).

Hence in the Diana model, the EBP is applied for each computed eigenfrequency/eigenmode considering both the energy introduced by the wind and the one dissipated by the system.

All the damping sources, such as spacers dampers and conductors are accounted for. For the eigenfrequencies and eigenmode where the amplitude of oscillations are not zero, the amplitude of motion and the resulting deformations are computed. Moreover, the model allows to account for frequency dependent spacer stiffness and damping and tension differential between bundle conductors.

Finally, the main outputs of the model consist in:

- The maximum amplitude of vibration over the whole span;
- The maximum amplitude of vibration registered in correspondence of spacer and damper clamps over the whole span;
- The strain in correspondence of singularities of the system: i.e. at the suspension clamp, at the armour rods and at the spacer and damper clamps;

It must be pointed out that the matrix transfer method, applied to the above described approach, has been also applied to the models developed in Russia for vibration of single (Afanasyeva et al. 1998a, b) and bundle conductors.

Claren–Cosmai Model

This model (Diana et al. 1982; Claren et al. 1971) is the same as the Diana model, using the same software. However, during the years some changes have been introduced in the formulation of the energy dissipated by the conductors and of the energy introduced by the wind.

4.3 Benchmarks

In this section, the results obtained for two different benchmarks are summarised: the first one is an analytical–analytical benchmark, i.e. the amplitudes and strains predicted by the various available models applied to a certain test case are compared; the second one is an analytical–experimental benchmark: in this case strains predicted by analysis are compared to strains actually measured in a field test in Pakistan. The experimental data were available to the benchmark participants since the beginning. In Table 4.2, the main information concerning the considered experimental tests are reported, while Table 4.3 summarises the spacer data. The input data of Tables 4.2 and 4.3 are used in both benchmarks.

In order to provide comparable results and highlight the main differences in the models, the same energy input from the wind and the same dissipated energy for the conductors have been introduced in the analytical–analytical benchmark (see Table 4.4).

Analytical–Analytical Benchmark Results

For the analytical–analytical benchmark the two considered models are the ones of Krispin and Diana, both of them use the same functions for wind power and conductor self-damping as shown in Table 4.4; it is then easier, from the results, to highlight the structural differences in the models.

The Claren–Cosmai model is not considered for this type of benchmark, because it should provide exactly the same results as the Diana model.

The predicted maximum vibration amplitudes for each mode of vibration are reported in Fig. 4.1, as a function of frequency, for the Diana and Krispin model. The predicted maximum strains are reported in Fig. 4.2.

Table 4.2 Experimental tests characteristics and system configuration data

Characteristics of the test span—Sect. 1 (Tarbela-Jalapur-Sharif)	
Location	Between towers 443–444 (susp.–susp.)
Span length (*l*) (m)	449
Terrain condition	Broken area, open terrain with low vegetation
Elevation above sea level (m)	494
Number of circuits	1
Phase configuration	Quad bundle, 457 mm separation
Phase conductor	
Type and name	ACSR Drake
Diameter (*D*) (mm)	28.14
Mass per unit length (*m*) (kg/m)	1.628
Ultimate Tensile Strength (UTS) (kN)	139.06
Tensile load (*T*) (kN)	32.0
Stranding	7 steel wires + 26 aluminium wires
Elementary wire diameters (mm)	3.45 (steel)–4.44 (aluminium)
Spacer damper	
Number	7
Subspan lengths (m)	37–63–55–68–61–69–59–37

Table 4.3 Spacer damper data

Central body mass	2.177	kg
Arm mass	0.735	kg
Central body moment of inertia	6.47×10^{-2}	kg m^2
Arm moment of inertia	1.57×10^{-3}	kg m^2
Hinge torsional stiffness KT	333	Nm/rad
Torsional loss factor	0.35	
Hinge axial stiffness KA	100	kN/m
Axial loss factor	0.2	

In both cases, the same tensile load is applied to the conductors of the bundle and a constant low wind turbulence is imposed, according to Tables 4.1 and 4.4.

In Fig. 4.1:

- the continuous lines, reported for reference purpose, show the results obtained in the case of a single conductor at the same tensile load of the bundle conductors (the blue line refers to Diana Model, the red one to Krispin model),
- the blue diamonds (Diana) and red crosses (Krispin) represent the maximum amplitude observed on the bundle conductors.

The diagram shows that, even if there are some differences in the two models, the trend in the results is very similar.

Table 4.4 Wind power input and conductor self-damping for the analytical–analytical benchmark

Wind power input	Conductor self-damping
	W/l = power dissipated per unit length of conductor (W/m) $W/l = DC\ U^2 f^4$ (4.1) where l: span length (m) U: vibration amplitude (m) f: vibration frequency (Hz) Conductor data: m: mass per unit length = 1.628 kg/m T: tensile load = 32000 N D: diameter = 0.02814 m DC: damping constant = 0.003089
Curves b, c, d are wind power input curves for one of the bundle conductors, for different turbulence levels The wind power input used in the analytical–analytical benchmark is the one represented by curve 'b', corresponding to low turbulence ($I_t < 0.07$)	Conductor self-damping used for the 'analytical-analytical' benchmark

As far as the maximum strain is considered, Fig. 4.2 shows the bundle maximum strain both in correspondence of the spacer clamp (green triangles and red diamonds) and the suspension clamp (violet triangles and blue diamonds), for Krispin (triangle points) and Diana models (diamonds points).

Also in this case, for the whole range of frequency considered, a generally good agreement between the results can be observed. There is a good agreement between the two model results for frequencies up to 25 Hz, while at higher frequencies, where the amplitude of vibration for both the models are very low (see Fig. 4.1), the Krispin model shows a level of strain higher than the one of Diana.

Analytical–Experimental Benchmark Results

In the analytical–experimental benchmark the models, described in Sect. 2, use for the wind power input and conductor self-damping their own selected function, as shown in Table 4.1.

Moreover, the data describing the field span and the spacer are the same as for the analytical–analytical benchmark (Tables 4.2 and 4.3).

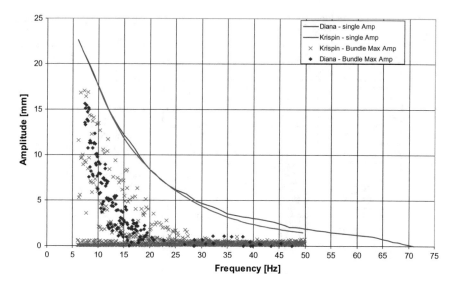

Fig. 4.1 Predicted maximum aeolian vibration antinode amplitude (0-peak) as a function of frequency when the same tensile load is applied on the subconductors and a constant low wind turbulence is considered

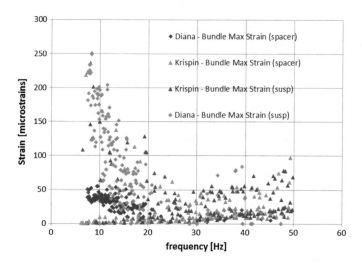

Fig. 4.2 Predicted maximum strain (0-peak) on the bundle conductors at the suspension clamp and at the spacer clamp as a function of frequency

The experimental results for the test case are available in the form of strains (0-peak value) at the suspension clamp, deduced by bending amplitude measurement performed through a typical vibration recorder.

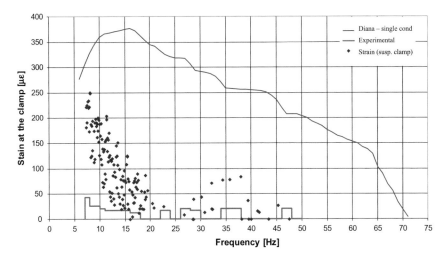

Fig. 4.3 Maximum strains at the suspension clamp using Diana model with the same tensile load applied to each subconductors of the bundle and a constant turbulence ($I_t < 0.07$)

Experimental and numerical results are compared, for the different models, in Figs. 4.3, 4.4, and 4.5 when the same tensile load is applied to each subconductor and a constant low turbulence is considered. The choice of a low turbulence level is related to the terrain description given in Table 4.2.

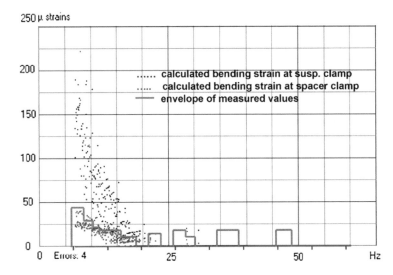

Fig. 4.4 Maximum strains at the suspension clamp using Claren–Cosmai model with the same tensile load is applied to each subconductors of the bundle and a constant turbulence ($I_t = 0.05$)

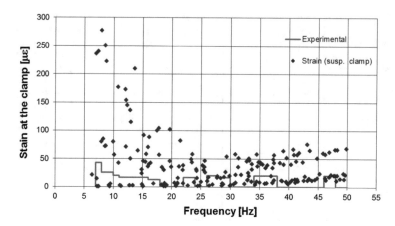

Fig. 4.5 Maximum strains at the suspension clamp using Krispin model with the same tensile load applied to each subconductors of the bundle and a constant, low wind turbulence

Figure 4.3 shows the maximum strain for the bundle conductors in the case of the Diana model. It can be seen that when the same tensile load is applied on the subconductors and a constant low wind turbulence index I_t lower than 0.07 is considered, the numerical data are conservative with respect to experiments in the frequency range 5–20 Hz.

In Fig. 4.4 the same analysis described for Fig. 4.3 is replicated using the Claren–Cosmai model. As expected, the results appear to be similar to the one obtained with the Diana Model.

The two considered turbulence levels ($I_t < 0.07$ and $I_t = 0.05$) have to be considered equivalent, representing the lowest turbulence levels present in each of the two computation programmes.

Finally, the same analyses as before are replicated by means of the Krispin model. As it can be seen, the numerical results are still conservative and similar to the previous ones (Fig. 4.5).

In Fig. 4.8 the maximum strains at the suspension clamp are reported when a 1D tension differential is applied to the bundle. The application of a tension differential always leads to a strain decrement in the bundle: it can be seen that the decrease of strain level when a 1D tension differential is applied is low.

Finally, in Fig. 4.9 the maximum strains at suspension clamps computed considering a tension differential of 10 D is reported. Moreover, in this calculation, the variation of stiffness and damping with respect to frequency for the spacer has been considered. It can be observed that in order to achieve a level of strain close to the experimental one, a tension differential of 10 D is required. This tension differential expressed in conductor diameters (10 D) corresponds to a percentage tension differential between upper and lower subconductors of the order of 2%.

The measured turbulence index (black points) in Fig. 4.6 is not that measured in the site to which the field measurements of the AE benchmark refer. It only

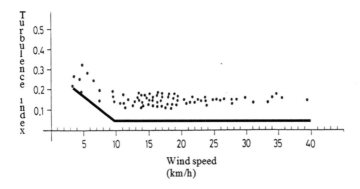

Fig. 4.6 Cosmai hypothesis for variable wind turbulence index

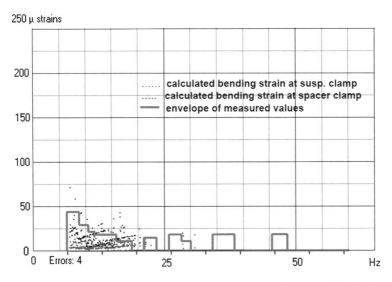

Fig. 4.7 Maximum strains at the suspension clamp using Claren–Cosmai model with the same tensile load applied to each subconductors and variable turbulence ($0.05 < I_t < 0.2$)

represents a typical trend of turbulence index, not related to a very flat terrain, showing that at low wind speeds, generally, turbulence increases as the wind speed decreases. Figure 4.7 shows analytical results obtained using this type of variable turbulence.

The influence of tension differential within the conductors of the bundle on aeolian vibration amplitude has already been investigated by Noiseux (Foata and Noiseux 1991). His conclusions have been reproduced using the Diana model (see Figs. 4.8 and 4.9).

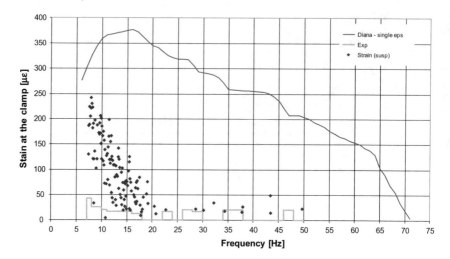

Fig. 4.8 Maximum strains at suspension clamp with Diana model: tension differential 1D and constant turbulence ($I_t < 0.07$)

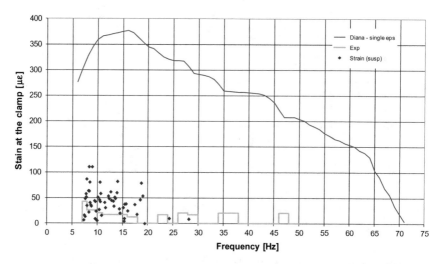

Fig. 4.9 Maximum strains at suspension clamp with Diana model: tension differential 10D and constant turbulence ($I_t < 0.07$)

4.4   Conclusions and Future Work

The two benchmarks assessed within the Working Group allowed to understand the main differences among the models presently adopted in the field of conductor vibrations to understand and control aeolian vibration.

The analytical–analytical benchmark showed that the computed vibration amplitudes have a very similar trend even if some differences in the models are present.

As far as the experimental–numerical benchmark, the numerical results generally exceed the experimental ones and then they are conservative, at least at low frequencies.

The benchmark has been developed on one case only, considering a quadruple bundle: this is, of course, a limitation and future work comparing analytical–experimental results from different cases (different bundle configurations and different conductors) should be planned.

In any case, it is needed to point out that, generally, when dealing with twin bundles, numerical results appear to be less conservative in respect to the experimental data (Diana et al. 1976, 1982).

The sensitivity analysis demonstrated that a non-negligible influence in the assessment of conductor behaviour, when dealing with aeolian vibrations, is given by the introduction of tension differentials and variable wind turbulence with wind speed.

Clearly, it is not straightforward knowing the real value to assign to the turbulence and to the tension differentials when the bundle behaviour for aeolian vibrations must be analysed.

Future work to achieve better knowledge on this issue should consider the comparison between measurements on a real line and analytical results.

Chapter 5
Modelling of Subspan Oscillations of Bundled Conductors

G. Diana, H. J. Krispin, J. Havel, J. Huang, J. Garnier,
S. Kolosov, J. P. Paradis, A. Leblond, C. B. Rawlins, U. Cosmai,
P. Van Dyke, J. L. Lilien, W. Troppauer, A. Manenti, L. Mazzola,
E. Ruggeri, N. Sahlani, D. Proctor, C. Rozé, M. Landeira, S. Thaddey,
C. Hardy, R. Tong, B. Liu, A. Bhangor, A. Vinogradov,
T. A. Furtado and M. Araujo

5.1 Introduction

Subspan oscillation is a well-known phenomenon in High Voltage and Ultra High Voltage Overhead Transmission Lines (HV and UHV OHTL) (EPRI 2009). It occurs on conductor bundles and it is due to the effect of the wake produced by the windward conductor on the leeward one. For this reason the phenomenon is also classified as wake-induced oscillations, this phenomenon is a flutter type instability due to the coupling of vertical, horizontal modes in a frequency range between 0.5 and 2 Hz.

Recently, problems associated with this phenomenon has become more recurrent, attracting the attention of transmission line operators, hence WG B2.46 decided to dedicate an effort to evaluate the present developments in such a field.

As for the analytical models developed for studying such a problem, the first ones are those of Simpson (1971), Ikegami et al. (1971), Diana and Giordano (1972), Ko (1973) and Tsui (1975). These are two degrees of freedom models: the motion of the leeward cylinder is studied along two orthogonal directions, the windward cylinder being still. The linearized quasi-steady theory (in the following QST) is employed and the drag and lift coefficients on the leeward cylinder are deduced from static measurements in wind tunnel, as a function of the relative position of the leeward cylinder with respect to the windward one. Such models are linear and clearly simplify the structural behaviour of the bundle sub-conductors, taking it back to a two degrees of freedom system in which the leeward conductor is the only one moving. Rawlins (1974, 1976 and 1977) expresses bundle dynamics in terms of normal propagation modes using the transfer matrix method. The bundle

G. Diana (✉)
Department of Mechanical Engineering, Politecnico di Milano, Milan, Italy
e-mail: giorgio.diana@polimi.it

© Springer International Publishing AG 2018
G. Diana (ed.), *Modelling of Vibrations of Overhead Line Conductors*,
CIGRE Green Books, https://doi.org/10.1007/978-3-319-72808-7_5

stability is analysed reproducing the aerodynamic forces through the QST with a linear approach. Nowadays the finite element model (in the following FEM) analysis allows for the reproduction of the bundle dynamics and for the application of the aerodynamic forces to each sub-conductor using the QST with a non-linear approach.

However, FEM analyses in the time domain are not always a practical tool for subspan oscillation simulation because of the computation time required to obtain results.

All the models developed until now rely on the quasi-steady theory (QST): according to this, the field of forces acting on the conductors in the wake is identified using the static aerodynamic coefficients measured in the wind tunnel and the effect of relative motion between sub-conductors is accounted for introducing a relative velocity with respect to the approaching flow (Diana and Gasparetto 1972; Diana et al. 1974). This approach generally holds for very high 'reduced velocities' V_r, defined as the ratio between the wind speed V and the product of oscillation frequency f and cylinder separation I: $V_r = V/(f\ I)$. Moreover, the problem being highly non-linear, the validity of this theory needs to be confirmed. Another important issue is the Reynolds number (Re) effect.

In fact for stranded conductors, i.e. rough cylinders, with the typical values of conductor's diameter and wind speed involved by subspan oscillations, Re may be close to the critical zone: hence the Re number could significantly affect the phenomenon, due to the non-negligible variations of the drag coefficient with Re itself (Simpson 1971; Wardlaw and Cooper 1973). As a matter of fact, recent researches made at Polimi (Diana et al. 2013) have shown that the QST maintains its validity for these values of 'reduced velocity'.

However, the same research shows evidence that the Reynolds number is very important because the conductors during subspan oscillations can operate around the critical range of the drag coefficient [see Fig. 5.1 (Wardlaw and Cooper 1973), where the drag coefficient of a 40.7 mm diameter Chukar conductor is reported, compared to that of a smooth cylinder with the same diameter]. For this reason, the static tests in wind tunnel to identify the aerodynamic coefficients on the leeward conductor must be made on rough cylinders, covering the critical and supercritical

Fig. 5.1 Drag coefficient on windward rough cylinder as function of speed/Reynolds number from Wind tunnel experimental tests (Wardlaw and Cooper 1973)

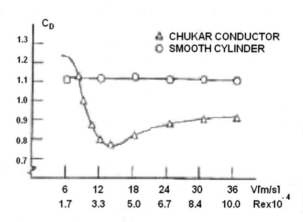

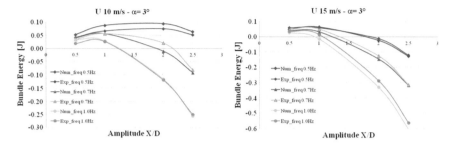

Fig. 5.2 Rough cylinders energy with respect to x/D amplitude for the three considered frequencies: experimental–numerical comparison. Due to the model scale, f = 0.5 Hz corresponds to f = 1 Hz full-scale α in both the static and dynamic bundle rotation angle with respect to the wind

range of the drag coefficient. Such an aspect has a direct impact on the evaluation of the bundle stability as demonstrated in Fig. 5.2, where the measured and computed energy on a couple of cylinders with one in the wake of the other is reported for different wind speeds, when the cylinders are moving along elliptical orbits (Diana et al. 2013). The energy input from the wind reported in Fig. 5.2 is a function of the subspan elliptical motion amplitude, represented on the horizontal axis in non-dimensional terms, as ratio x/D between the amplitude and the conductor diameter.

As can be seen, increasing the vibration amplitude, the wind energy input is decreasing, becoming negative, so the subspan phenomenon is self-controlled. The Reynolds effect makes the bundle less stable at low speed, as can be seen comparing the energy input at 10 m/s to that at 15 m/s in Fig. 5.2 (please note that, to held readability, the vertical scales are different in the two Figures).

Additionally, it can be observed that the higher the frequency of the motion, the more stable the bundle appears, hence it is confirmed that a beneficial effect on subspan oscillation can be obtained by decreasing the subspan length.

In the present work, a benchmark within the different type of models at disposal for subspan oscillation studies is carried out comparing numerical results with measurements on the IREQ Varennes test line equipped with a quad bundle of ACSR Bersimis conductors and spacer dampers.

5.2 Analytical Methods and Assumptions Used in the Study

In this section a brief description of the analytical methods adopted in the study is provided, together with the assumptions introduced in the analysis.

Diana Model

The Diana Model is an energy-based method that first evaluates the natural frequencies and vibration modes of the bundles, then identifies the modes showing a predominant horizontal and vertical component and selects those in the frequency range 0.5–3 Hz, frequency range typical of subspan. Subsequently within the selected modes, the ones which could be coupled by the aerodynamic forces to give rise to subspan oscillations are chosen.

For each possible pair of horizontal and vertical modes, the possibility of having instability is evaluated by a quasi-steady theory (QST) linear approach and the range of wind speeds for which subspan oscillation may be excited together with an instability index are computed. This preliminary analysis provides also the maximum difference between the vertical mode frequency and the horizontal mode frequency allowing for the coupling between the two modes.

Once the types of modes that can be coupled are defined, two independent modal coordinates $q_o(t)$ for the horizontal mode and $q_v(t)$ for the vertical mode of the bundle are chosen. Harmonic laws are imposed to the two coordinates $q_o(t)$ and $q_v(t)$, in such a way to reproduce the elliptical motion of the conductors typical of flutter instability. Being the maximum instability index found for a phase shift between the two modes of $\pm\pi/2$, the imposed law is

$$
\begin{aligned}
q_0(t) &= S_{am} * e^{i\omega t} \\
q_v(t) &= S_{im} * e^{i\left(\omega t \pm \frac{\pi}{2}\right)}
\end{aligned}
\tag{5.1}
$$

where S_{am} and S_{im} represent the maximum amplitudes for the horizontal and vertical modes, being the modes normalized in such a way that the maximum amplitude of the cylinder motion along the whole span is equal to unit, and ω is chosen as the circular frequency of the horizontal mode. Finally, the conductor's motion all along the span is obtained multiplying $q_o(t)$ and $q_v(t)$ by the modal shapes.

Using the QST, the forces acting on the conductors in the different positions occupied along the elliptical trajectories are computed, taking into account the velocity of each conductor (Diana and Gasparetto 1972; Diana et al. 1974).

Once the aerodynamic forces on the conductors are known, it is possible to compute, in each section of the bundle, the energy introduced by the forces themselves in one complete elliptical cycle.

The method gives the energy introduced by the wind in the entire bundle, summing up the energies computed in each section: the evaluation is performed both for $+\pi/2$ and $-\pi/2$ phase shift [see Eq. (5.1)], obviously choosing the situation for which the aerodynamic forces can introduce maximum energy into the bundle system.

Finally, the steady-state amplitudes of oscillation are defined through the balance between the energy introduced by the aerodynamic forces and the energy dissipated by the bundle, equipped with the spacer dampers.

The Reynolds effect plays an important role in the drag and lift aerodynamic coefficients of the conductors, both for the windward and the leeward ones. This effect is introduced in the model thanks to the results of a recent research conducted in the wind tunnel of Polimi (Diana et al. 2013).

Sergey and Vinogradov Model

The Sergey and Vinogradov method (1986, 2003) relies on the direct numerical simulations of the conductors and spacer dampers movement in the process of subspan vibrations. It is fully based on the finite differences method and the explicit scheme of system equations solution. The equations of movement initially compiled in 3D form, for the sake of simplicity, are reduced to a 2D problem taking into account negligible longitudinal displacements of spacer dampers along the span length. The calculation of the bending stresses in the conductors is performed assuming that the spacer damper clamp divides conductor's bending angles uniformly. The aerodynamic coefficients are taken after Diana and Giordano (1972) and Price and Païdoussis (1984). The assumption that Poffenberger–Swart formula is valid while subspan oscillation process is taking place is made. Moreover, the model has been demonstrated to be valid for quad bundle span, where in order to replicate properly the subspan oscillation, an initial condition in terms of force field is applied on the bundle: two short startup impulses of the forces acting in opposite directions are applied on a pair of interacting conductors. In case of even number (e.g. four) of conductors in the bundle, it turned out most effective to take into interaction only one pair of conductors. The solutions taken prevent the model to find only the snaking motion and allow to detect subspan oscillations. But lack of wind energy influx provided lower sensitivity of the model to weak and moderate winds; this disadvantage was corrected by a coefficient found with the help of the benchmark test itself.

Snegovskiy and Lilien Model

The corresponding model relies on the use of finite element non-linear formulation. Within this approach, the interaction of sub-conductors due to the wake is represented using Simpson's aeroelastic model (Simpson 1971). A special force element is created to introduce the aerodynamic loads due to the wake which are computed according to the QST. Moreover, the aeroelastic properties of the wake force field are tuned to meet the wake-induced instability properties, as measured by Price and Païdoussis (1984). Extension of the wake interaction sample onto the full line span (including spacer damper, any bundle configuration) is done by taking into account the inertia–stiffness properties of the line fittings (spacer dampers). More details are given in (Snegovskiy and Lilien 2010; Snegovskiy 2010).

Kurmann Model

The idea of the model approach is a finite element implicit transient analysis with an update of the aerodynamic forces after each time step. The geometry model consists of linear beam elements for the conductor, the frame and the arms of the spacer damper. The stiffness and damping properties of the spacer damper hinges are considered as spring–damper elements. The self-damping of the conductor is implemented with the Rayleigh damping model.

The first two load steps for the pretension of the conductor and the dead load are executed without time integration. Afterwards, a do-loop with time integration and update of the aerodynamic forces is performed. For the numerical solution the HHT (Hilber–Hughes–Taylor) algorithm is applied.

The quasi-steady theory (QST) is used for the computation of the aerodynamic forces. The resultant polynomials of the wind tunnel test according to Diana and Gasparetto (1972) are the input for the model approach. The latest wind tunnel test results (Diana et al. 2013) are also included, which quantify the influence of the Reynolds number onto the aerodynamic coefficients.

5.3   Experimental and Analytical Benchmark: Description and Results

It must be pointed out right from the beginning that the models using FEM can give rise, at wind velocities between 10 and 20 m/s, as reported in detail in (Snegovskiy and Lilien 2010), to an instability motion of the whole span at very low frequency also called galloping or snaking motion. This instability motion is predominant and this approach is generally not able to reproduce subspan oscillation. However, a full description of the results obtained by FEM model is reported in (Snegovskiy and Lilien 2010).

To overcome this problem and allow for the excitation of subspan oscillation, the Sergey and Vinogradov model uses, as initial condition, an external excitation which moves the sub-conductors out of phase.

By means of this approach the model is able to reproduce subspan oscillation in a reasonably short time (after 30 s of real time process development).

In order to compare the model's accuracy and reliability, an experimental–numerical benchmark has been carried out, considering as a reference case the one tested in Varennes by IREQ, which is described in Table 5.1.

The compared quantities are the maximum peak-to-peak oscillations of the sub-conductors in the most critical sections of the span, during subspan oscillations.

The experimental benchmark was obtained on the Hydro-Quebec full-scale test line which has two dead-end spans and three suspension spans. The length of the spans is successively 150, 400, 450, 425 and 150 m. The subspan oscillations were measured on the middle span (450 m span) (Fig. 5.3).

The subspan oscillations were measured in the middle of each subspan on the bottom North–East conductor to obtain the horizontal component of the first sub-span mode–antinode amplitude. The data was measured during 4 weeks every ten minutes at a rate of 32.3 pts/s for a duration of 145.5 s. The time signal was then processed to determine the peak-to-peak amplitude of the most severe vibration

Table 5.1 Subspan oscillation test case data

Spacer damper data	
Quad spacer damper interaxis (mm)	457
Torsional stiffness of the hinge (Nm/rad)	125
Radial stiffness of the hinge (N/m)	50,000
Ratio between torsional damping and torsional stiffness	0.35
Ratio between radial damping and radial stiffness	0.35
Arm mass (kg)	0.71
Arm mass moment of inertia with respect to the centre of mass (kg m^2)	2.74×10^{-3}
Central body mass (kg)	3.44
Central body mass moment of inertia with respect to the centre of mass (kg m^2)	38.5×10^{-3}
Mass of spacer damper (kg)	6.28
Span data	
Conductor	ACSR Bersimis
Diameter (mm)	35.1
Mass (kg/m)	2.185
Tensile load (N)	34,420
Span length (m)	450
Subspan arrangement (m)	40-53-57-50-55-49-58-52-36

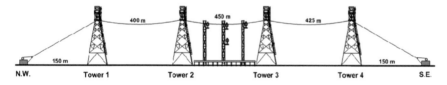

Fig. 5.3 Test setup showing towers, conductors and anemometers of the test line

cycle during the recording period and the vibration frequency. Only the recordings with an apparent frequency corresponding to the first vibration mode of the con- sidered subspan and/or of its adjacent subspans were selected. The purpose here was to distinguish between subspan oscillations appearing in the above-mentioned frequency range from snaking oscillations, which usually occur at lower frequen- cies, and also from rain vibrations or aeolian vibrations, which have higher fre- quencies. Figure 5.4 reports the maximum peak-to-peak conductor amplitudes measured in the different subspans.

The numerical simulations have been carried out in the speed range 20–60 km/h.

Figure 5.5 shows, as a function of mean wind speed, the peak-to-peak horizontal oscillation amplitude registered in subspan 7, which is found to be the most critical one.

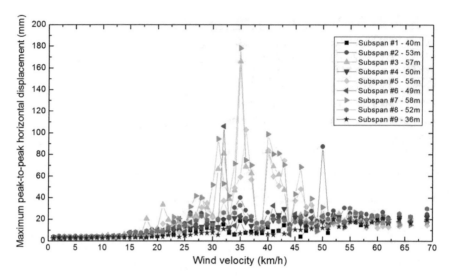

Fig. 5.4 IREQ Measurements: Maximum peak-to-peak horizontal oscillations as function of mean wind speed

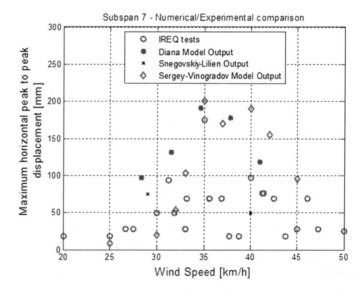

Fig. 5.5 Benchmark results: Maximum peak-to-peak horizontal oscillations as a function of mean wind speed

In blue the results from the Diana model are reported, in green the results from Sergey–Vinogradov model and in black the results from Snegovskiy and Lilien model, while in red the experimental points are represented.

It is possible to observe that the Diana and Sergey and Vinogradov numerical models correctly replicate the trend of the oscillation peak-to-peak values, which increase with the increase of mean wind speed up to reach a maximum in correspondence of a mean wind speed of 35 km/h, then the oscillation amplitude starts to decrease.

The numerically computed amplitudes are also in good agreement with the ones experimentally registered.

5.4 Conclusions

The study presented several approaches to the evaluation of the subspan phenomenon, ranging from approaches based on the EBP to approaches relying on FEM modelling.

Subspan oscillations is a complex vibration on bundle conductors. It needs relatively high winds. Important parameters are:

- Bundle tilt
- Ratio vertical to horizontal frequency in each subspan
- Tensile load
- Ratio between bundle separation I and conductor diameter D (I/D)
- Subspan length.

Simple and advanced methods can be used nowadays. Complex FEM model have been applied with success but they need a very cumbersome analysis which provides results that depend strongly on details that are not easy to quantify.

Methods based on modal analysis and energy approach seem to be a more useful tool for practical applications.

Moreover the presented results, even if applied to only one test case, allow to state that:

- The quasi-steady theory seems to be able to well-reproduce the aerodynamic forces produced during subspan oscillations.
- The Reynolds number affects in a large amount the energy introduced by the wind. In fact, the flow on the conductors can vary from subcritical, critical and supercritical region of Reynolds number depending on the conductor diameter, surface roughness and finally, wind speed. In the critical region the energy introduced is greater due to the decrease of aerodynamic damping on the upstream cylinder associated to the negative slope of the drag coefficient. It can be observed that, considering two conductors with the same diameter but different surface roughness, the critical region moves towards higher wind speeds for smoother conductors, like those with trapezoidal wires with respect to the standard conductors with circular strands. This aspect of the problem is for sure worth further investigation through suitable wind tunnel tests.

- The numerical model based on EBP approach and on sophisticated wind tunnel tests to identify the aerodynamic parameters seems a very useful tool for analysing the subspan oscillations phenomenon, as shown by the comparison between its results and the measurements on a full-scale transmission line structure subjected to the real wind.
- The numerical model based on the finite differences method and the explicit scheme of system equations solution, even if it had to be calibrated with the help of the benchmark test itself, also seems in condition to well-reproduce the phenomenon. However, more results and analyses are required to confirm its validity.

From the analyses performed in this research and, more precisely, from the results shown in Fig. 5.2, it is possible to conclude that one way of controlling subspan oscillations is to increase the subspan oscillation frequency by decreasing the subspan length.

General Conclusions

This brochure has highlighted what can be expected from numerical models regarding conductor vibrations. In the first chapter, the aeolian vibrations of single conductors are obtained using the EBP and compared with experimental results. If the range of uncertainty of the wind power functions and self-damping models which are the main input of those models is indicative of the range of uncertainty of the predicted amplitudes, then the range of uncertainty in EBP predictions of vibration amplitude is about ±50–60%.

Generally, the highest vibration strains are found at relatively low frequencies, where the contribution of dampers to overall damping is much more important than that of conductor self-damping. This is not true at high frequencies. Thus, the largest deviations between models for self-damping, which occur at the lowest frequencies are less important than they may appear.

In practice, assessment of the aeolian vibration condition of particular lines, with conductors whose mechanical properties are poorly defined, or with special terrain conditions, may require field measurements. Techniques to perform such measurements have already been described [CIGRE SC22 WG11 TF2, 1995].

The same exercise has been performed on single conductors equipped with one damper at the end of the span. Benchmark results show a wide dispersion of the predicted maximum amplitude values, and large discrepancies between the analytical predictions and the measured behaviour of the test span. On the other hand, while the dispersion of the dynamic stiffness of different damper samples was reflected in corresponding dispersion in predicted amplitudes, it is not a reflection on the technology, since this dispersion will remain the same, independently of the technology used. Nevertheless, analytical methods based on the EBP and shaker-based technology can provide a useful tool for use in the design of damping systems for the protection of single conductors against aeolian vibrations. It should be used with circumspection and be supplemented by references to field experience.

The modelling of aeolian vibrations has also been evaluated for very long spans strung at relatively high tensions with multi-damper arrangements. It is more challenging since the wind velocity is not homogeneous along such long spans. Furthermore, tension at the support can be significantly greater than at mid-span due

© Springer International Publishing AG 2018
G. Diana (ed.), *Modelling of Vibrations of Overhead Line Conductors*,
CIGRE Green Books, https://doi.org/10.1007/978-3-319-72808-7

to the large sag involved. This can cause the coupling between the vortex-induced forces and the conductor, as well as self-damping, to vary along the span.

The details of the calculation results revealed large differences in predicted clamp amplitudes among the dampers under consideration during steady vibration at a constant frequency, for most of the frequencies covered. Furthermore, these predicted clamp amplitudes were in general very different from the amplitudes on which the damper characteristics were based and, as already observed, represent one of the reasons for the discrepancies between experimental and analytical results.

Future research work is needed to improve the EBP technology, which generally produces a safe design of the damping system.

On bundles aeolian vibrations, the numerical results generally exceed the experimental ones, consequently, they are conservative, at least at low frequencies, however, the comparison was made only with a quad bundle. Other published work (Diana et al. 1976, 1982) has shown that, generally, when dealing with twin bundles, numerical results appear to be less conservative with respect to experimental data.

The sensitivity analysis demonstrated that a non-negligible influence in the assessment of conductor behaviour, when dealing with aeolian vibrations, is given by the introduction of tension differentials and variable wind turbulence with wind speed. Clearly, it is not straightforward knowing the real value to assign to the turbulence and to the tension differentials when the bundle behaviour for aeolian vibrations must be analysed.

Four models have been presented to evaluate the subspan oscillation phenomenon. FEM modelling has been applied with success but it needs a very cumbersome analysis which provides results that depend strongly on details that are not easy to quantify. Methods based on the modal analysis and energy approach seem a more useful tool for practical applications.

As previously shown experimentally, the following parameters have a predominant impact on subspan oscillation amplitudes: bundle tilt, the ratio of vertical to the horizontal frequency in each subspan, tensile load, the ratio of bundle separation over conductor diameter and subspan length. Moreover, the flow on the conductors can vary from sub-critical, critical and super-critical depending on the Reynolds number thus on the conductor diameter, wind speed and surface roughness (conventional vs. trapezoidal) and it has a paramount influence on the energy introduced by the wind.

There was only one experimental benchmark and more results are required to validate the models, however, modelling has shown that one way of controlling subspan oscillations is to increase the subspan oscillation frequency decreasing the subspan length.

This brochure has reported the state of the art regarding aeolian vibrations and subspan oscillations modelling. Of course, this field of expertise is not static and research, as much numerical than experimental, is still going on in order to improve our knowledge of the phenomenon and translate it into improved numerical models.

References

Afanasyeva, O., Ryzhov, S., Feldshteyn, V.: Dynamic models for the study of vibration Power Line conductors and communication cables in the air flow, Problems of Mechanical Engineering and Automation (1) 50–57 (1998a)

Afanasyeva, O., Feldshteyn, V., Ryzhov, S.: Calculation of rational geometry of the helical support clamp under the terms of vibration. Electricheskie Stantsii (Power Plants) (1), 12–17 (1998b)

Anderson, K., Hagedorn, P.: On the energy dissipation in spacer-dampers in bundled conductor of overhead transmission lines. J. Sound Vib. **180**(4), 539–556 (1995)

Bate, E.: The vibration of transmission line conductors. Transaction of the Institution of Engineers (Australia), vol. X, pp. 277–290 (1930)

Belloli, M., Resta, F., Rocchi, D., Zasso A.: Wind tunnel investigation on aeroelastic behaviour of rigidly coupled cylinders. In: Proceeding 5th International Symposium on Cable Dynamics, Santa Margarita Liguria (Italy), 15–18 Sept 2003 (2003)

Brika, D., Laneville, A.: Laboratory investigation of the power imparted by wind to a conductor using a flexible model. In: IEEE (P.E.S.) Winter Meeting (1995)

Brika, D., Laneville, A.: The power imparted by wind to a flexible conductor in the wake of another. IEEE Power Eng. Rev. **17**(1) (1997)

Carrol, J.S.: Laboratory studies of conductor vibration. Electr. Eng. Trans. AIEE Power Apparatus Syst. **55** (1956)

Cigada, A., Manenti, A.: Sulla misura dello smorzamento proprio dei conduttori con il metodo dell'onda stazionaria. Rapporta Interna del Dipartimento di Meccanica del Politecnico di Milano 8–96 (1996)

CIGRE SC 22: WG on Mechanical Oscillations Guide on Conductor Self-Damping Measurements. Guide pour les mesures d'auto-amortissement des conducteurs. Electra no. 62, pp. 79–90 (1979)

CIGRE SC 22 WG 01: Report on aeolian vibration (Rapport sur les vibrations éoliennes). Electra no. 124, pp. 41–77 (1989)

CIGRE SC 22, WG 11, TF 2: Guide to vibration measurement on overhead lines (Guide des mesures des vibrations sur les lignes aériennes). Electra no. 163

CIGRE TF 22.11.01: Modelling of aeolian vibration of single conductors: assessment of the technology. Electra no. 181, pp. 52–69 (1998)

CIGRE TF B2.11.04: Overhead conductor safe design tension with respect to aeolian vibrations. Technical Brochure 273 (2005)

CIGRE TF B2.11.01: Modelling of aeolian vibrations of a single conductor plus damper: assessment of technology. Electra no. 223, pp. 28–36 (2005)

CIGRE WG B2.31: Modelling of aeolian vibration of single conductors strung at relatively high tensile load. Application to HV & UHV lines. Electra no. 256, pp. 34–40 (2011)

CIGRE WG B2.46: Wind induced motion on bundled conductors (excluding ice galloping) Part A —Aeolian vibrations (To be published in 2015)

© Springer International Publishing AG 2018   73
G. Diana (ed.), *Modelling of Vibrations of Overhead Line Conductors*,
CIGRE Green Books, https://doi.org/10.1007/978-3-319-72808-7

CIGRE WG B2.46: Wind induced motion on bundled conductors (excluding ice galloping) Part B
—Subspan oscillations (To be published in 2015)

Claren, R., Diana, G.: Mathematical analysis of transmission line vibration. IEEE Trans. Power
Apparatus Syst. **PAS-88**(12) (1969)

Claren, R., Diana, G.: Mathematical analysis of transmission line vibration–IEEE Trans 31C83 '67

Claren, R., Diana, G., Giordana, F., Massa, E.: The vibrations of transmission line conductor
bundles. In: 71 TP 158 PWR, IEEE Winter Meeting (1971)

Diana, G., Cheli, F., Fossati, F., Manenti, A.: Aeolian vibrations of overhead transmission lines:
computations in turbulence conditions. J. Wind Eng. Industrial Aerodyn. **46 & 47**, 639–648 (1993)

Diana, G., Di Giacomo, G., Claren, R.: An approach to vortex shedding under turbulent air flow on
a single vibrating cylinder. In: IEEE 1979 Summer Meeting, New York, 15 July (1979)

Diana, G., Falco, M.: On the forces transmitted to a vibrating cylinder by a blowing fluid
(experimental study and analysis of the phenomenon). Meccanica **VII** (1) (1971)

Diana, G., Falco, M., Gasparetto, M.: On vibrations due to vortex shedding induced on two
cylinders with one in the wake of the other. Meccanica (3) (1976)

Diana, G., Falco, M., Cigada, A., Manenti, A.: On the measurement of overhead transmission lines
conductor self-damping. IEEE Trans. Power Delivery **15**(1), 285–292 (2000)

Diana, G., Giordana, F.: Sulle vibrazioni di un cilindro in scia di un altro – Analisi del fenomeno
con modello analitico. L'Energia Elettrica (7) pp. 448–457 (1972)

Diana, G., Gasparetto, M.: Energy method for computing the amplitude of vibration of conductor
bundles due to wake effect. L'Energia Elettrica (1972)

Diana, G., Gasparetto, M., Di Giacomo, G., Nicolini, P.: Analytical method for computing subspan
oscillation Analytical and Experimental results. In: C 74 493-3, IEEE Summer Meeting,
Anaheim California, USA, 14–19 July 1974 (1974)

Diana, G., Gasparetto, M., Tavano, F., Cosmai, U.: Field measurement and field data processing
on conductor vibration (comparison between experimental and analytical results). In: CIGRE'
International Conference on Large High Voltage Electric Systems, 1982 Session (1982)

Diana, G., Manenti, A., Cigada, A., Belloli, M., Vanali, M., Pirotta, C., Zuin, A.: Stockbridge type
damper effectiveness evaluation: parts I and II. IEEE Trans. Power Delivery **18**(4), 1462–1477,
Oct 2003 (2003)

Diana, G., Manenti, A., Belloli, M., Muggiasca, S., Bousseau, P., Peyrard, C., Relun, N.:
Modeling of Aeolian vibration of overhead line conductors with wind speed variations along
and across the span. In: 8th International Symposium on Cable Dynamics, Clamart (France),
20–23 Sept 2009 (2009)

Diana, G., Belloli, M., Manenti, A., Muggiasca, S., Guglielmini, S., Bousseau, P.: Impact of
turbulence on vortex induced vibrations and fatigue of conductors: modelling and real span
experimentation. In: CIGRE General Session, Paris 2010, Paper B2-303 (2010)

Diana, G., Belloli, M., Giappino, S., Manenti, A., Mazzola, L., Muggiasca, S., Zuin, A.: Wind
Tunnel testing developments in the last 50 years. In: 2013 European and African Conference
on Wind Engineering, Robinson College, Cambridge, UK, 7–11 July 2013 (2013)

EPRI: Transmission line reference book—wind-induced conductor motion, 2nd. edn, pp. 3–25.
Electric Power Research Institute, Palo Alto, Ca 2009 1018554, USA (2009)

Falco, M., Gasparetto, M., Nicolini, N., Di Giacomo, G.: Comportamento vibratorio di lunghe
campate di attraversamento, equipaggiate con conduttore singolo e con dispositivi smorzanti.
L'Energia Elettrica (5) 311–324 (1973)

Farquharson, F.B., McHugh, R.E.: Wind tunnel investigation of conductor vibration using rigid
models. Trans. AIEE Power Apparatus Syst. **75** (part III) (1956)

Foata, M., Noiseux, D.: Computer analysis of asymmetries of a two-conductor bundle upon its
aeolian vibration. IEEE Trans. Power Delivery **6**(3) (1991)

Hadulla, T.: Vortex-induced vibrations in overhead line conductor bundles. Doctoral Thesis,
Technische Universität Darmstadt (2000)

Hagedorn, P.: On the computation of damped wind-excited vibrations of overhead transmission
lines. J. Sound. Vib. **83**(2), 263–271(1982)

Hagedorn, P., Schmidt, J.: The effect of a damper on a vibrating conductor calculated by use of the complex mechanical impedance of the damper, Annals Cigre 22-86 (WG01) 05, Nov 1985 (1985)

Hardy, C.: Comparison of theoretical calculations and field measurements on single conductors. CIGRE 22-91 (WG11-TF1) 9 (1991)

Hardy, C., Leblond, A.: Comparison of conductor self-damping measurements. CIGRE SC 22-93 (WG 11) 88 (1993)

Hardy, C., Van Dyke, P.: Field observations on wind-induced conductor motions. J. Fluids Struct. 43–60 (1995)

IEC 61897: Requirements and tests for stockbridge type aeolian vibration dampers. (1998)

IEEE Std. 563-1978: Guide on conductor self-damping measurements (1978)

IEEE Std. 664-1993: Guide on the measurement of the performance of Aeolian vibration dampers for single conductors (1993)

Ikegami, R., et al.: Structural analysis. High Voltage Power Transmission Systems—Boeing Co. Seattle BPA Ctr 14-03-1362N (1971)

Ko, R.G.: Theoretical investigation for Hydro-Quebec into the aerodynamic stability of bundled power line conductors, Part I: two dimensional stability analysis of a conductor in the wake of a fixed conductor. Lan Tech Rep LA-122, NRC (Canada) (1973)

Kraus, M., Hagedorn, P.: Aeolian vibrations: wind energy input evaluated from measurements on an energized transmission line. IEEE Trans. Power Delivery 6, 1264–1270 (1991)

Leblond, A., Hardy, C.: Aeolian vibration calculations of conductor-damper systems, 22-00 (WG11-TF1) 09, Kolding, May 2000 (2000)

Leblond, A., Hardy, C.: On the computation of safe design tension with regard to aeolian vibration of damped single conductors. In: Proceeding of the 4th International Symposium on Cable Dynamics, pp. 137–144, Montréal, 28–30 May 2001 (2001)

Leblond, A., Hardy, C.: On the estimation of a 2x2 complex stiffness matrix of symmetric stockbridge-type dampers. 3rd International Symposium on Cable Dynamics, Trondheim, Norway, Aug 1999 (1999)

Möcks, L., Schmidt, J.: Survey of measurements of mechanical self-damping of ACSR conductors. CIGRE SC 22-89 (WG 11) TFI-2 (1989)

Noiseux, D.U.: Similarity laws of the internal damping of stranded cables in transverse vibrations. IEEE Trans. Power Delivery 7(3), 1574–1581 (1992)

Noiseux, D.U., Houle, S., Beauchemin, R.: Transformation of wind tunnel data on aeolian vibration for application to random conductor vibrations in a turbulent wind. IEEE Trans. Power Delivery 83(2), 263–271 (1988)

Palazuelos, E., Izquierdo, J., Fernandez, A.: Control of conductors to prevent construction problems. In: 1989 International Conference on Overhead Line Design and Construction: Theory and Practice, 84–88 (1988)

Papailiou, K.O.: On the bending stiffness of transmission line conductors. IEEE Trans. Power Delivery 12(4) Oct 1997 (1997)

Pon, C.J., Havard, D.G., et al.: Aeolan vibrations of bundle conductors. Canadian Electrical Association, Report No. 177 T 510 (1989)

Price, S.J., Païdoussis, M.P.: The aerodynamic forces acting on groups of two and three circular cylinders when subject to a cross-flow. J. Wind Eng. Ind. Aerodyn. 17, 329–347 (1984)

Rawlins, C.B.: Recent developments in conductor vibration research. Technical Paper No. 13, ALCOA Lab (1958)

Rawlins, C.B.: Effect of wind turbulence in wake–induced oscillations of bundled conductors. In: Conference Paper C74 444–6, IEEE PES Summer Meeting, Anaheim, CA, 14–19 July 1974 (1974)

Rawlins, C.B.: Fundamental concepts in the analysis of wake-induced oscillation of bundled conductors. IEEE Trans. Power Apparatus Syst. PAS–95(4), 1377–1393 (1976)

Rawlins, C.B.: Extended analysis of wake-induced oscillation of bundled conductors. IEEE Trans. Power Apparatus Syst. PAS–96(5), 1681–1689 (1977)

Rawlins, C.B.: Power imparted by wind to a model of a vibrating conductor. Technical Report No. 93-82-1 EL Prod. Div, ALCOA Lab., Massena, N.Y. (1982)

Rawlins, C.B.: Model of power imparted by turbulent wind to vibrating conductor. Report No. 93-83-3, Alcoa Conductor Products Co., Spartanburg, South Carolina (1983a)

Rawlins, C.B.: Notes on the measurements of conductor self-damping. Report W 93-83-4, Alcoa Lab., Alcoa Conductor Products Co. (1983b)

Rawlins, C.B.: Wind tunnel measurements of the power imparted to a model of vibrating conductor. IEEE Trans. Power Apparatus Syst. **PAS-102**(4) (1983c)

Rawlins, C.B.: An efficient method for measuring dissipation by dampers in laboratory spans. IEEE Trans. Power Delivery **3**(3) (1988)

Rawlins, C.: Conductor self-damping data and comparison between theoretical and experimental data on aeolian vibration on single conductor. CIGRE 22-92 (WG11-TF1) 12 (1992)

Rawlins, C.B.: Calculation of damping efficiency of dampers from damper impedance. 22-97 (WG11-TF1) 31, Sendai. Oct 1997 (1997)

Rawlins, C.B.: The long span problem in the analysis of conductor vibration damping. IEEE Trans. Power Deliv. **15**(2) (2000)

Seppa, T.: Self-damping measurements and energy balance of ACSR drake. IEEE WPM, Paper No. 71 CP-161 (1971)

Sergey, I., Vinogradov, A.: Mathematical simulation and calculation experiment for different forms of conductors dynamics. In: Proceeding 5th International Symposium on Cable dynamics, pp. 93–100, Santa Margherita, Italy, 15–18 Sept 2003 (2003)

Seveso, A.: Politecnico di Milano, Dipartimento di Meccanica Valutazione sperimentale dello smorzamento strutturale del conduttori e proposta di una metodologia di misura da applicare al metodo dell'onda stazionaria. Tesi Laurea di G. De Nardis, A. Seveso, AA. 95–96 (1996)

Simpson, A.: On the flutter of a smooth circular cylinder in a wake. Aeronaut. Q (1971)

Slethei, T. O., & Huse, J.: Conductor vibration—theoretical and experimental investigations on a laboratory test span. In: Proceedings of the Institution of Electrical Engineers, IET Journals & Magazinese, 112(6), 1173–1179 (1965)

Snegovskiy, D.: Wake-induced oscillations in cable structures: finite element approach. Available on http://bictel.ulg.ac.be/ETD-db/collection/available/ULgetd-07142010-102207/ (2010)

Snegovskiy, D., Lilien, J.L.: Nonlinear finite element approach to simulate wake-induced oscillation in transmission system. In: Proceedings of ASME 2010 3rd Joint US-european Fluids Engineering (FEDSM2010). Available on http://hdl.handle.net/2268/102157 (2010)

Streliuk, M.I., Serguey, I.I. Vinogradov, A.A., Krasnov, V.A.: Dynamic loads on spacers at subspan oscillations of the EHV T/L conductors. Energ. Stroit. (1), 67–71 (in Russian) (1986)

Tavano, F.: Results of self-damping measurements on conductors for overhead lines and on earth wires with optical fibres. Cigre SC 22-88 (WG 11) 18 (1988)

Tavano, F.: Collection of experimental data on aeolian vibration on single conductors. CIGRE 22-91 (WG11 TF1) 6 (1991)

Tompkins, J.S., Merrill, L.L., Jones, B.L.: Quantitative relationships in conductor vibration damping. IEEE Trans. PES **75**, 879–894 (1956)

Tsui, Y.T.: Two dimensional stability analysis of a circular conductor in the wake of another. IEEE Paper A75 576-9 (1975)

Wardlaw, R.L., Cooper, K.R.: A wind tunnel investigation of the steady aerodynamic forces on smooth and stranded twin bundled power conductors for the Aluminum Company of America. Lab Tech Rep LA-117, NRC (Canada) (1973)

Printed in the United States
By Bookmasters